AF342356

THE ENVIRONMENT
PROBLEMS AND SOLUTIONS

A Collection of New Studies and Outstanding
Dissertations on Current Issues

STUART BRUCHEY

Allan Nevins Professor Emeritus
American Economic History
Columbia University
GENERAL EDITOR

A GARLAND SERIES

GROUNDWATER PROTECTION FROM PESTICIDES

BRIAN BAKER

Garland Publishing, Inc.
NEW YORK & LONDON 1990

Library of Congress Cataloging-in-Publication Data

Baker, Brian.
Groundwater protection from pesticides / Brian Baker.
p. cm. — (The Enviromnment—problems and solutions)
Includes bibliographical references.
ISBN 0-8240-2568-7 (alk. paper)
1. Pesticides—Environmental aspects—New York (State)—Suffolk County.
2. Water, Underground—Pollution—New York (State)—Suffolk County. 3.
Pesticides—Government policy—New York (State)—Suffolk County. 4.
Pesticides—Economic aspects—New York (State)—Suffolk County. I. Title.
II. Series.
TD427.P35B35 1990
363.73'84—dc20   90-47329

CONTENTS

TABLES

FOREWORD

The environmental consequences of modern agriculture have become increasingly apparent. Intensive plant crop production and concentrated animal operations threaten natural resources in a way that was unthinkable a century before. One resource that has come under pressure has been groundwater. Pesticide contamination has been recently discovered in many agricultural areas. This study examines the institutions that fail to protect this critical resource.

This study began as a doctoral dissertation. Since the original work was filed, I have been able to expand on the original analysis and refine the economic model. Much of the model appeared originally in the *Northeastern Journal of Agricultural Economics*, and appears here by their permission. The conclusion reflects both the hope and frustration of research on alternatives that promise to reduce the environmental impact of agriculture. In particular, the lack of availability of budgets and research on organic farming methods caused these methods to be excluded from analysis. Understanding of biological and cultural pest controls lag so far behind chemical control methods that it is difficult to predict their value. However. if agriculture is to continue in environmentally sensitive areas, new technologies must be compatible with environmental conditions.

PREFACE

Groundwater contamination by agricultural practices presents a dilemma between protecting a vital resource and preserving a valuable part of the economy. This book critically evaluates the institutions developed to protect drinking water from pesticide contamination: water pollution control, pesticide registration, drinking water quality standards. These structures fail to solve the cause of the problem.

Long Island is used as a case study. This unique region presents a case where groundwater contamination problems are particularly troublesome. The dominant crop of the region is potatoes. Monocultural practices have led to severe pest problems, including a large infestation of the Colorado potato beetle. Methods of pest control have employed large doses of pesticides. The hydrogeology of the region makes the underlying aquifer particularly susceptible to pesticide leaching. The resulting contamination has caused two pesticides to be banned, and a third to be withdrawn from the market. Producers have had difficulty adjusting to the unavailability of these means to reduce Colorado potato beetle damage.

The author constructs a model that integrates the biological, economic and geochemical situation of agriculture in Eastern Suffolk County. This model consists of a recursive programming component, which has input for it generated by a model of Colorado potato beetle pest dynamics and management strategies to control those pests, and an index of contamination. The model is run under different policy settings, which include no regulation, taxation of pesticides, ban of selected pesticides, forced crop rotations, taxation of potatoes, purchase of crop rights, ban of potatoes, and the development of pest control districts. The remainder of these policy settings take as given the ban on pesticides. While income is reduced by banning pesticides, the reduction is small when compared with the improvement in environmental quality. Further efforts to reduce pesticide use resulted in a reduction in potato acreage and incomes, as well as yields.

While short-run economic considerations favor the status quo, a long-run perspective encourages further efforts to reduce pesticide use and coordinate economic and environmental considerations.

ACKNOWLEDGEMENTS

There are too many people to thank individually for this book, but I'd like to acknowledge some of the more important contributors. Foremost is David Allee, who was a constant source of encouragement from the beginning. His patience and wealth of knowledge allowed me to develop a rich research plan. Richard Boisvert tirelessly critiqued the work, and saw to it that the final product was of high quality. Fred Aman contributed valuable insights of the legal problems that face the institutions charged with resource management, and was very receptive to the interface between law and economics. George Fohner, Gerry White, Sheryl Lazarus and Mildred Warner all permitted me to borrow from their intellectual creations. Dave Newton, Bill Sanok and Dale Moyer gave me access to Suffolk County's agriculture. Dave Andow, Sandy Galaini, Dave Lansky, Maurie Semel, Christine Shoemaker and Bob Wright assisted me in understanding pest population dynamics and the biology of the Colorado potato beetle. Taamo Steenhuis, Chris McVoy, Hank Hughes and Steve Pacenka all answered questions about groundwater dynamics and modelling. Dan Bartholomew and Ben Schwarz both helped me out of many programming jams. Peter Berck and David Zilberman critiqued the structure of my model, and showed me how to streamline both the symbols and text of my work. Portions of this work appeared in the *Northeastern Journal of Agricultural and Resource Economcis*, and appears by the kind permission of Editor Loren Tauer. Richard Norgaard added his knowledge of resource management in the later drafts. Scott Bentley and especially Chuck Bartlet helped assemble the finished text. Finally, David Kay and Ralph Gifford showed me that it's how you play the game that counts.

GROUNDWATER
PROTECTION
FROM PESTICIDES

CHAPTER I

Pesticides in Groundwater

Groundwater contamination by agricultural chemicals presents a challenge to the food production system. Techniques used to supply food threaten irreversible contaminating of a vital resource. Can modern technology adapt to provide both abundant food and a clean environment? The answer lies in the institutions that direct technology and the incentives producers face in choosing technology.

THE PROBLEM

Farmers have polluted groundwater for centuries. Wells have been polluted by manure and fouled by buried livestock. These incidents were relatively localized, with the polluters being adversely affected the most. While understanding of hydrology was relatively primitive, there was a great incentive to not foul one's own source of fresh, pure water. This limited both the extent to which groundwater was threatened, and the economic dimension of the pollution, known as externalities.

Modern agricultural technology has changed farming, rural America and the environment. Increased use of pesticides has resulted in accelerated degradation of the groundwater resource.[1] Of particular concern are acutely toxic and carcinogenic pesticides.[2] Pesticides have been found in groundwater all across the United States, from Maine to California.[3]

The mechanism by which chemicals leach into the water table is only now beginning to be understood. The ability of chemicals to reach the saturated zone of the soil depends on the chemical's molecular

3

structure, soil structure, temperature, whether foliar or soil application, and a host of other factors.[4]

The problem is complicated on the suburban/rural fringe, where agricultural practices often come in conflict with residential land use. The exposure to pesticides in groundwater affects more than farmers. Farmland has been traditionally viewed in suburban areas as a means to preserve open space. With the environmental impact created by pesticide contamination, many suburban communities are questioning agriculture's place in the environment.

The problem is often viewed as the need to resolve the contradiction between opposing goals: the preservation of agriculture and the protection of groundwater. This study maintains that, while there is no single best policy to achieve optimum levels of each, improvements can be made through effective public policy. To implement effective policies, institutions managing land and water resources must be restructured. Most importantly, they must take into account the interconnections between land use, water quality, agricultural practices and ecological impact. Policies that deal with these different areas need to be coordinated.

Institutions to prevent or abate the pollution of groundwater are inadequate for dealing with agricultural sources. The common law of nuisance and negligence is able to compensate victims of pollution after the fact. Because of the time required for groundwater to purify itself, however, relief in the form of clean water may be impractical, if not impossible. Federal statutes which cover the problem of agricultural pollution of groundwater include: the Federal Water Pollution Control Act, the Safe Drinking Water Act and the Federal Environmental Pollution Control Act. Not one of these specifically addresses the issue of agricultural pollution of groundwater. States have also enacted legislation which regulates water pollution, controls pesticide use, sets drinking water standards and encourages improved management of agricultural land.[5] States may be reluctant to implement programs which would improve groundwater quality at the expense of farm income.

Local regulations have also been introduced to prevent farm nuisances and protect water supply sources. These ordinances may be inadequate because they fail to cover a broad enough geographical area to protect an entire recharge zone for an aquifer, or local governments may not have adequate resources. Local governments do have a strong incentive and some administrative advantages. However, local ordinances are subject to preemption by state and federal authority.

Drinking water standards are a parameter given by the law. The cost of meeting these standards is ordinarily borne by the water suppliers and passed on, where possible, to the consumers in the water bill. This is because standards are usually met by treating the water supply. One alternative to treatment is source protection. If the objective of a public water supplier is to provide abundant amounts of water which meet health and safety standards as cheaply as possible, source protection may prove to be a viable alternative if the cost of treatment is high relative to the cost of protection.

Most systems have either minimal protection and full-scale treatment facilities, or minimal treatment and stringent source protection. Some systems have both measures at a sophisticated level; many rural systems have a minimal amount of treatment and virtually no source protection. The amount of treatment that a community has is subject to both its willingness and ability to pay. Add to these inadequacies the conflicting objectives of preserving prime agricultural land. Difficult choices need to be made to decide which is of greater importance: unpolluted groundwater or a strong farm sector. The tradeoff must account for long-run as well as short-run needs, because in the long-run, the two goals may converge. Clean water is needed for productive agriculture.

RESEARCH OBJECTIVES

The primary objective of this research is to predict the effect of different policy on the use of and eventual contamination of groundwater by pesticide application on agricultural crops. To do this, it is necessary to construct a model of the behavior of pesticide users. Different possible policies are introduced into the model to see how pesticide use changes in response to the policies and economic incentives created by institutional change. The effects of these policies on the incomes of farmers and their decisions on land use are also examined. Once the results of these policies are attained, the amount of pesticides that reach the groundwater can be predicted.

To analyze the effect of different policies on these two different objectives, it is necessary to understand the effect of policy on behavior which controls agriculture and the activities which pollute groundwater. This information can be used to examine the tradeoff between the quality of groundwater and the economic viability of agriculture. Another objective is to examine the institutional support for various strategies for groundwater protection. The institutions which implement a policy are likely to dictate the success or failure of a policy. All too often, a policy is designed with little thought as to the way it is to be delivered.

SUMMARY AND METHODOLOGY

Chapter 2 begins with a treatment of the general problems which agriculture causes its neighbors. The reader is given a brief description of agricultural pollution problems. This is followed by a summary of existing policies and legislation dealing with agricultural pollution and pesticide use. The ways in which current law and policy fail to adequately address groundwater contamination by agriculture are shown. At the end of that chapter some alternatives to deal with the shortcomings of existing policy are proposed. The structure of existing institutions which deal with groundwater management is critically examined. The administrative, regulatory and financial factors for managing agricultural pollution of water are critiqued. This critique shall serve as a point of departure for proposing alternatives.

Pesticide technology and regulations are critically examined in chapter 3. Chapter 4 explores the role of standards in making environmental policy. Different paradigms for standard setting are set forward. The policy for setting standards is evaluated in terms of each of these paradigms: Technocratic rationality, bounded rationality, pluralistic bargaining and speculative augmentation.

To apply these diverse elements to an existing situation of agricultural pollution, a region where pesticides have contaminated groundwater is studied. The region chosen for this thesis is Suffolk County, on Long Island. Because of the unique nature of the situation on Long Island, and Suffolk County's agriculture, chapter 5 is devoted to the events which led to the current problems.

The basic framework for analysis is presented in chapter 6, where a model simulating Long Island's agriculture is built. Concepts in linear and recursive programming are briefly reviewed. The components of the model are presented. The economic component consists of a regional recursive programming model. The use of pesticides is modelled by an economic model which has a pest population dynamics component. The output of the model is designed to be used as input for a groundwater transport model. Existing groundwater models are considered.

Chapter 7 explains the various policies, including both their legal and economic implications. The model is run over a period of five years to predict the performance of the different programs over time. The results of the economic model and the final results will be interpreted to compare the effectiveness of the different policies in how well they balance economic and environmental values. The policies reviewed include *laissez faire*; the ban of pesticides found in the groundwater; forced crop rotation; a tax on pesticides; a subsidy for low-input crops; and the implementation of a pest control district. The use of the output in a groundwater model is presented.

While no single criteria for evaluating policy is developed, the often conflicting objectives of different policies are taken into account in the analysis of the different results. Current policy, which prohibits the use of certain pesticides that have appeared in the groundwater, are found to be an effective second-best solution for both economic and environmental criteria. Results suggest that environmental quality can be improved by collective action to control pest populations at relatively little cost. However, the incentives must be properly structured to encourage voluntary cooperation and a minimum of disruption to the farm sector.

Agricultural pollution of groundwater has long been a problem, but that problem has become more acute as technology changes and population is concentrated. The laws and institutions to deal with the problem have undergone rapid development in the past decade. One crucial aspect of the problem is the interconnection between land and water resources. This has been the weak link of the institutions charged with the management of groundwater. To better understand this link, let us look at the ways in which society has attempted to deal with the problem of agricultural pollution.

NOTES
CHAPTER I

1. A good summary of the literature is contained in George R. Hallberg, "Overview of Agricultural Chemicals in Ground Water," *Agricultural Impacts on Ground Water--A Conference.* (Dublin, OH: Water Well Journal Publishing Co.: 1986): 1-66.

2. V. I. Pye, R. Patrick and J. Quarles, *Groundwater Contamination in the United States* (Philadelphia: University of Pennsylvania Press, 1983).

3. US Environmental Protection Agency, *National Survey of Pesticides in Drinking Water Wells.* (Washington: US Environmental Protection Agency, 1986); Elizabeth G. Nielsen and Linda K. Lee, "The Magnitude and Costs of Groundwater Contamination from Agricultural Chemicals: A National Perspective." (Washington, DC: US Department of Agriculture, Economic Research Service, 1987); Deborah M. Fairchild, "A National Assessment of Ground Water Contamination from Pesticides and Fertilizers," in Fairchild (ed.) *Ground Water Quality and Agricultural Practices.* (Chelsea, MI: Lewis Publishers, 1987): 273-294; Joe D. Francis, Bruce L. Brower and Wendy F. Graham, "National Statistical Assessment of Rural Water Conditions." (Washington: US Environmental Protection Agency, 1982);

4. See R. Allan Freeze and John Cherry, *Groundwater.* (Englewood Cliffs, NJ: Prentice-Hall, 1979) for a text on groundwater movement.

5. See US Congress, Office of Technology Assessment, *Protecting the Nation's Groundwater from Contamination.* (Washington: OTA Reports, 1984);

CHAPTER II

Agricultural Pollution of Groundwater

AGRICULTURAL POLLUTION

Pollution resulting from standard agricultural activities has created conflict between two objectives that seek to preserve environmental amenities. Agriculture, through pesticide use and soluble fertilizer application, and by large scale animal operations, threatens the integrity of groundwater resources in many parts of the nation. On the other hand, agriculture provides amenities through open space.

Agriculture has diminished in economic importance, farm population and cultivated land as society has become more urbanized. This has been made possible by technological advances that have increased agricultural sector productivity and freed labor for non-farm employment. This same technology also interfere with neighboring property owners' rights. The keeping of animals and spreading of manure can cause odors.[1] Spraying and dusting pesticides may cause property damage.[2]

There are several ways in which law and public policy recognizes the problem described above. Common law has been used, to a varying degree. any of these activities could be the basis for civil action against a farmer. Numerous nuisance and negligence suits for these causes have been filed against farmers by neighbors. Federal and state statutes have also sought to alleviate the problem. However, no law directly deals with the specific problem, so coverage is incomplete.

In hearing a case brought against a farmer charging him with nuisance, the court must balance the gravity of the harm with the

9

utility of his conduct in assessing whether or not the farm he operates is a nuisance. If the harm is great from farming practices, this would outweigh the utility of farm production.[3] The suitability of the area for agriculture is another important factor. This test requires that farming be compatible with other uses.[4]

As rural non-farm population grows and the value of competing land uses increase, the more likely agriculture is to be found as incompatible with surrounding uses. One defense that farmers have employed against nuisance suits has been the doctrine of 'coming to the nuisance.' The fact the farm was there first can be a deciding factor in a suit brought against a farmer, with the case being decided in the farmer's favor.[5] Being there first is not an invincible defense.[6] As farming operations become larger, the potential for farms to create nuisances also increases. This is especially true of livestock raising operations. The odor emanating from concentrated feedlots has added a new dimension to an old problem. The law developed to manage agricultural pollution problems in the past has not been able to remedy the groundwater pollution problems created by modern technology. Common law and Federal statutory responses are reviewed.

COMMON LAW APPROACHES

The very nature of nonpoint pollution makes it difficult to bring civil action against nonpoint polluters. A long, convoluted, inconsistent history of common law governing water pollution has been extensively documented.[7] Common law doctrines that govern water pollution revolve around whether the system is appropriative or riparian. If riparian, the principle question is whether the undiminished flow rule or the reasonable use rule is followed. To quote Davis,[8]

> . . . riparians have a legal right to discharge waste, provided the discharge is reasonable with respect to other riparian uses and provided a private or public nuisance is not created or maintained.

The court would base its decision, then on 1) the reasonableness of pollution with respect to other uses and 2) whether or not the pollution creates a nuisance. This differs from the view of Prosser who states,[9]

> There is no riparian right or privilege to pollute water, nor
> do landowners have rights to pollute surface and ground
> water on their land.

The court clearly recognizes no right to pollute where it interferes with the property rights of owners of water rights.[10]

> It cannot be denied that a well of water is property
> recognized by the law, any injury to which is redressible
> by law. To pollute or foul the water of a well is an
> actionable injury.

Common law offers two theories that the courts use to abate water pollution: negligence and nuisance. A survey of cases brought against polluters of groundwater showed the two to have been utilized about equally in percolating groundwater pollution situations.[11] Two other theories which have been used less frequently are strict liability and the special rule proposed in Prosser's *Restatement of Torts*.[12] The negligence theory governing groundwater pollution requires that a groundwater user knows or should have known that his or her activity would be likely to cause the injury which has occurred.[13]

For groundwater pollution to be considered a private nuisance, the pollution must interfere with private property interests.[14] The two grounds are not mutually exclusive. A number of nuisance suits over the pollution of water have involved farmers as both plaintiffs and defendants.[15] Farmers have sued neighboring farmers for negligence in polluting agricultural water supplies.[16] Cases where farmers sue farmers for agricultural pollution indicate two things: that some forms of pollution are beyond the scope of normal farming activities, and that farmers are often victims of agricultural pollution.

The ability of common law to remedy water pollution has been criticized. Common law is inadequate to deal with the complex problem of groundwater contamination.[17] The courts have failed to recognize both the changes in technology that affect groundwater contamination and improved understanding of hydrology.[18] Common law negligence and nuisance actions are ineffective because certain communities depend on groundwater as their sole water supply.

Groundwater recharges at a slow rate. Once an aquifer is contaminated, it is likely to remain so for an extended period of time. While it is technically feasible to restore the quality of a contaminated aquifer, the cost of doing so is so great that few courts would order it as relief. Prudence to prevent pollution costs less than remedies to restore an aquifer to its pristine state. The common law was considered sufficiently weak to merit augmentation by the key statute enacted in the 1970's to remedy water pollution: The Federal Water Pollution Control Act.[19]

REGULATORY APPROACHES

EPA has been delegated the authority to distinguish between point and nonpoint source pollution.[20] Agricultural pollution control is generally considered non-point. Enforcement of non-point standards have caused intergovernmental conflicts. The objectives and authority of the Federal, state and local levels have taken different roles in regulating, protecting and promoting agriculture.

Regulations that force agriculture to abate pollution would have a different effect on farms than they would on other enterprises. Farms, unlike firms in many manufacturing industries, lack the market power to pass the costs of pollution on to the consumer. Farmers are price takers, and as such must absorb cost increases. There is a reluctance among officials for this and other reasons to impose more stringent regulations on farmers.[21]

When Congress passed the Federal Water Pollution Control Act (FWPCA) in 1972,[22] EPA was given, for the first time, direct enforcement power over water pollution. While attention was given to nonpoint sources of pollution, point sources were subjected to stricter standards and scrutiny than were nonpoint sources. This dilemma is well documented in the context of animal feedlot regulations.[23] After the passage of the FWPCA, EPA was charged with the responsibility of drafting regulations which, among other things, restricted point source pollution from "concentrated animal feeding operations."[24]

EPA submitted regulations that would have required virtually every farmer with a feedlot to have a permit for waste-water discharge.[25] After facing much opposition from farm groups, EPA relaxed the regulations considerably.[26] These laxer regulations were challenged in *Natural Resources Defense Council v. Train.*[27] The court held that EPA was not authorized under the FWPCA to exempt point sources of pollution from the permit process.[28] Quoting the NRDC, the court saw the permit process as necessary for the enforcement of the FWPCA.[29] However, the court recognized the great burden of requiring permits of every feedlot owner in the country,[30] and noted that Congress apparently delegated the authority to EPA to define point and nonpoint pollution.[31] Given that administrative discretion, EPA promulgated regulations that were essentially similar to those challenged in *Train.*[32]

Section 208 of the FWPCA provides for the establishment of regional plans for the development of water and related land resources.[33] This planning was supposed to innovate state management of nonpoint pollution. The planning process has not fulfilled the original, ambitious expectations spelled out in the FWPCA. The ability to control nonpoint pollution through 208 regional plans has disappointed Congress.[34] Among the reasons for the lack of success in formulating these plans include funding and administrative problems, and inadequate water quality data.[35] EPA never had a clear idea how the 208 process was to work and changed its philosophy several times over the course of its administration. Section 208 planning may also have been hampered by a lack of expertise in relating land use controls to water quality.[36] The lack of clear direction, lack of data and lack of expertise have led to shortcomings in the program.

The most directly relevant part of section 208 for agriculture is the Culver amendment.[37] The Culver amendment directs the Secretary of Agriculture to establish and administer Best Management Practices (BMP) of agricultural and silvicultural lands.[38] These practices are carried out by farmers who are under contract with the Department of Agriculture. The responsibility for monitoring the farmers rests on the Soil Conservation Districts and the Soil Conservation Service. The Secretary of Agriculture is authorized to provide cost sharing for BMPs that qualify.

The main intention of BMPs was to control surface water pollution from soil runoff. Like the rest of the FWPCA, the Culver amendment failed to explicitly address groundwater pollution problems. BMPs can be implemented in a way that reduces groundwater pollution, but are inadequate for dealing with serious and

extensive contamination problems.[39] A separate set of regulations deal
exclusively with pesticides.

SUPERFUND

The Comprehensive Environmental Response, Compensation and
Liability Act, also known as CERCLA or Superfund was originally
intended to clean up massive environmental problems caused by
hazardous waste disposal. CERCLA may be increasingly used to
correct pesticide contamination problems. The addition of six wells
drawing from groundwater contaminated by normal use of pesticides
to EPA's priority list for cleanup action in October, 1984 has given
CERCLA a role in pesticide contamination problems.[40] This action has
been controversial, prompting some to challenge EPA's authority to
do so under the act.[41] Even if EPA has the authority to use CERCLA
for pesticide contamination cleanup, one can't help but wonder if EPA
would be willing to do so. The expense of cleaning up hazardous waste
already exceeds the amount earmarked by Superfund. The additional
expense of responding to pesticide contamination incidents may be
viewed as a drain on resources intended for other purposes.

The main function of the Federal government is to provide
funding, technical assistance, research and development.[42] Pye, Patrick
and Quarles state,[43]

> Although there is no one single statute specifically aimed at
> protecting groundwater, there are more than a half dozen federal
> environmental statutes which, in combination with each other,
> are directed at accomplishing that objective. Most significant of
> these are [the Resource Conservation and Recovery Act, Safe
> Drinking Water Act, Federal Water Pollution Control Act, and
> the Comprehensive Environmental Response, Compensation
> and Liability Act]. A review of these and other environmental
> statutes indicate that there is a legal framework which can
> significantly aid in the implementation of a groundwater
> protection strategy.

The variable conditions faced by different states and communities
requires that a national groundwater policy be able to respond to and
accommodate a wide range of problems. The site specific nature of

groundwater contamination requires flexibility in policy administration.

Office of Technology Assessment,[44] and others, have proposed a Federal Groundwater Act that would coordinate these various programs. Such an act would have to balance the needs of the arid parts of the country with those of the humid, populous areas. Some changes, such as those affecting the use of interstate aquifers, would attract vehement special interest support or opposition. While there is certainly a need to augment and coordinate existing legislation, it is unlikely to overcome legislative inertia.

The Office of Groundwater Protection in EPA was created to coordinate existing laws. There is some difference of opinion about what, if any, role the Federal government should play in groundwater protection. While the Federal statutes mentioned above touch upon groundwater law, there really is no comprehensive Federal authority over groundwater. Most of the responsibility and authority has been delegated to state and local government.

STATE AND LOCAL GOVERNMENT

Groundwater protection techniques vary from state to state, and even within states. The Environmental Law Institute called these techniques the nuts and bolts of groundwater quality management.[45] ELI explored five different specific protection techniques: Groundwater quality standards, source oriented controls, groundwater user regulations, land use controls, and containment or cleanup requirements. Each of these has advantages and disadvantages. Numerical standards set a maximum level of contamination by specific chemicals. Standards have several advantages and disadvantages.[46]   Allowable levels of contamination are clearly defined and administrative discretion is removed.   Standards are explicitly set before proceedings against polluters are initiated. Dischargers, users and disposers are influenced by a uniform set of standards.

The disadvantages of standards are that they are expensive and time consuming to set for the vast number of contaminants. Standards do not prevent contamination by themselves. Once the maximum level is exceeded, it is too late to prevent contamination. Setting standards requires a complex understanding of risk assessment and expertise not available to many states. Standards can lead to rigid, and expensive control technologies for dischargers. Once standards are set, they are difficult to change, even in light of new scientific information. The problems of implementing standards will be dealt with in greater detail in chapter 4.

There are many types of source oriented controls, but only two are readily applied to agricultural non-point pollution: performance standards establishing zones of discharge and best management practices. Connecticut, Florida is one state that has adopted a zone of discharge scheme.[47] Connecticut's program relies on dischargers purchasing easements to cover the land area necessary for treatment. This is applied only to sewage and other biological wastes--manure in the case of agriculture. Soils are relatively ineffective in filtering pesticides. Florida's zone of discharge system is more complicated, resembling land use controls. Discharges beyond 100 feet of the site of discharge or within 100 feet of the property line are prohibited. Discharges overlying a potable water supply are restricted to stormwater and sewage, all other discharges are prohibited. Less protective groundwater classes allow additional types of discharges, but any discharges which pose an imminent hazard to drinking water supplies are prohibited. Discharge zones are more flexible than land use controls, but are more performance oriented than standards.

Best Management Practices (BMPs) are devised to minimize the potential for particular activities to contaminate groundwater. BMPs are cost-effective compared to other means, and can be implemented on a site-specific basis. However, they are limited in their ability to reduce pollution because they are difficult to mandate and enforce. They can require extensive management and coordination by administrative agencies. If Section 208 is any indication of the results one could expect with groundwater management, the results would be inadequate. BMPs may be more effective if they are used in cross-compliance with Federal subsidy programs.

Regulation of groundwater use is a more common technique to protect groundwater from agricultural pollution in states which have a great deal of irrigated agriculture. Groundwater management in these situations. Four techniques are described in the ELI report: Licensing well drillers, well regulations, withdrawal or use permits, and return flow/recharge requirement.

ELI sees land use controls as most suited for restricting activities on identifiable, sensitive areas. While they have the potential to be very effective, land use controls are also highly controversial. Land use controls examined by ELI include zoning regulations; siting, development and construction regulations; public acquisition programs; and transferable development rights. Zoning and building regulations are more directed at residential, commercial and industrial sources of pollution, but can also be used for agriculture. Acquisition of water rights is an option more readily exercised in states under the appropriative doctrine. TDRs have not been widely used.[49]

Most state legislation aimed at controlling nonpoint source pollution has resulted from the requirements section 208 of the FWPCA. Most notable of these for agriculture are the Soil and Water Conservation Districts.

In New York State, Soil Conservation Districts were formed in 1940.[50] A 1975 amendment to the Soil and Water Conservation District Law was designed to meet the Federal 208 requirements for water and related land use planning.[51] Soil and water conservation districts have primarily limited themselves to investing in capital improvements to reduce erosion. The county government is responsible for the creation and funding of soil conservation[52] and small watershed districts,[53] and for funding agriculturally oriented activities such as cooperative extension.[54] However, soil and water conservation and small watershed districts have been given no power to enact land use controls, such as zoning. The districts have no regulatory function per se, and as such they rely heavily on farmer cooperation, voluntary participation, and support from the U.S. Department of Agriculture's Soil Conservation Service, and Agricultural Stabilization and Conservation Service. While they have had success at reducing sedimentation, reducing other forms of agricultural pollution is beyond their grasp.

States have complained that there is a lack of adequate Federal assistance to implement state programs mandated by Federal legislation. Many states expressed the need for greater funding for BMP programs, and the creation of a Federal clearinghouse for information on state and local legislation to protect groundwater.[55]

Local ordinances were the first type of regulatory control used to protect water supply sources. These had their origins in the late 1800's when water borne diseases, such as typhoid, diphtheria and cholera were common problems in the U.S. As the technology to prevent and cure water borne diseases improved, watershed rules and regulations were less relied upon.[56] Chemical technology changed the nature of water supply contamination in two ways. First, chemical treatment virtually eliminated pathogens from public water supplies. Second, evolving and widespread use of chemicals in industry and agriculture has caused a wide range of toxic pollutants to enter water supplies. These chemicals have adverse health effects, most notably cancer. They can enter the water supplies in a number of ways. Water supply source protection in New York State has not changed much since the 1940s.[57] area is readily identified, as with large aquifers underlying porous soils.[58]

State aquifer classification systems have been used to manage and protect groundwater. A classification system defines which aquifers are in need of protection and which are not. New York State has an extensive groundwater classification system. Regulations prohibit discharges which impair the best use of groundwater sources or which are "deleterious, harmful, detrimental or injurious to the public health, safety or welfare."[59] Aquifers are divided into three classifications: Class GA, which are the most stringently protected; class GSA; and class GSB, the least stringently protected.[60] Protection of the groundwater is the responsibility of the state Department of Environmental Conservation. The quality drinking water is handled by a separate agency, the state Department of Health. This has, at times, made the implementation of groundwater and drinking water protection awkward. To further add to the problems with the state program, NYCRR specifically excludes any "normally accepted agricultural practices of utilizing chemicals and fertilizers for the growing of crops for human and animal consumption" from the effluent standards for discharges to class GA waters.[61] This severely restricts regulation of agricultural practices.

California has extensive groundwater contamination problems.[62] Recent legislation has sought to abate pesticide contamination of groundwater in two ways. The first has been to collect data on the potential for pesticides to leach. Those likely to leach or with significant data gaps on leachability will have their registrations suspended.[63] The state is required to carry on an extensive groundwater monitoring program.[64] If pesticides continue to be found in the groundwater, they can have their registration removed.[65] The other is the Safe Drinking Water and Toxic Enforcement Act of 1986.[66] This act states that "No person in the course of doing business shall knowingly discharge or release a chemical known to the state to cause cancer or reproductive toxicity into water or onto or into land where such a chemical passes or probably will pass into any source of drinking water . . . "[67] Prop. 65 works through liability and civil penalties.[68] It also provides the opportunity for public interest lawsuits.[69]

Given the strong resistance to land use controls in general and increased regulation of farming practices in particular, it is unlikely that the groundwater protection through land use controls will be rapidly implemented. Other alternatives, such as regulation of pesticides and standard setting for drinking water, must also be considered. If groundwater itself is not directly protected from agricultural pollutants, further environmental damage can be prevented by regulating the contaminants. This is precisely the aim Federal and state laws governing the use of pesticides.

NOTES

CHAPTER II

1. *Siegle v. Bromley* 20 Colo. 189, 124 P. 191 (1912); *Mercer v. Brown*, 190 So.2d 610 (Fla. 1966); *Gerrish v. Wishbone Farms of New Hampshire*, 231 A.2d 622 (N.H. 1967); *Bower v. Hog Builders*, 462 S.W.2d 761 (Mo. 1970); *Meat Producers Inc. v. McFarland*, 476 S.W.2d 845 (Tex. 1972); *Patz v. Farmegg Inc.* 196 N.W.2d. 557 (Ia. 1972); *Botsch v. Leigh Land Co.* 195 Neb. 54, 236 N.W.2d 815 (1975).

2. *A. Gerrard Co. v. Fricker* 42 Ariz. 503, 27 P.2d 678 (1933).

3. Prosser *Restatement (2d) Torts* §§827, 828, 829.

4. Prosser, *Restatement (2d) Torts.* §§828(b), 831.

5. *Walker v. Wearb*, 6 N.Y.S.2d 548 (1938); *Michelsen v. Leskowicz*, 55 N.Y.S.2d 831, affirmed 63 NYS2d 191.

6. *Spur Industries, Inc. v. Del E. Webb.* 108 Ariz. 178, 494 P.2d 700 (1972).

7. Peter N. Davis, "Theories of Water Pollution Litigation", 1971 *U. Wis. L. Rev.* 738 (hereinafter cited as Water Pollution Litigation); Peter Junger, "Bad Water: Common Law Nuisance and Welfare Economics Mixed Well", 27 *Case West. Res. L. Rev.* 3 (1976); Peter N. Davis, "Groundwater Pollution: Case Law Theories for Relief", 39 *Mo. L. Rev.* 117 (hereinafter cited as Groundwater Pollution); Note, "Private Remedies for Water Pollution", *Columbia Law Review* 70: 734.

8. Davis, *Water Pollution Litigation*, p. 780.

9. *Restatement (2d) Torts*, 849(e) (1970).

10. *U.S. v. Alexander*, 148 U.S. 186, 191-192 (1892).

11. *Groundwater Pollution* p. 125.

12. *Restatement of Torts*, (1939), §822.

13. *Phillips v. Sun Oil Co.*, 307 NY 328, 331, 121 N.E.2d 249, 251 (1954); *Collins v. Chartiers Valley Gas Co.*, 131 Pa. 143, 159, 18 A. 1012, 1013 (1890).

14. Prosser, *Restatement (2d) Torts* section 821D (1970).

15. *Atkinson v. Herington Cattle Co.*, 200 Kan. 298, 436 P.2d 816 (1968), feedlot runoff polluted a well used by a dairy farm; *Spence v. McDonough*, 77 Iowa 460, 42 N.W. 371 (1889), hog pond made water unfit for cattle; *McEvoy v. Taylor*, 56 Wash. 357, 105 P. 851 (1909), farm animals polluted livestock water supply;

16. *Randall v. Clifford*, 119 Vt. 216, 122 A.2d 833 (1956), buried cow polluted dairy farm's water supply; *Van Brocklin v. Gudema*, 50 Ill. App. 2d 20, 199 N.E.2d 457 (1964), manure pile contaminated livestock well.

17. Davis, *Groundwater Pollution*, 144-146; Junger, supra, 3.

18. Corker, *Groundwater Law, Management and Administration*, pp. 112-126.

19. P.L. 92-500, 33 U.S.C. sections 1251-1376 (1982).

20. 40 Fed. Reg. 54183 (1975).

21. Richard Booth and Albert Bronson, *Major Institutional Arrangements Affecting Groundwater in New York State*, (Ithaca, Center for Environmental Research, 1983).

22. P.L. 92-500, 33 U.S.C. 1251-1376.

23. John E. Montgomery, "Control of Agricultural Water Pollution: A Continuing Regulatory Dilemma." 1976 *Univ. of Ill. Law Forum* 533; Recker, *Sw. J. of Law*; Hines, "Farmers, Feedlots and Federalism" 19 *S.D. Law R.*

24. 33 U.S.C. 1362(14).

25. 38 Fed. Reg. 18000 (1973).

26. See 39 Fed. Reg. 506 (1974).

27. 396 F. Supp. 1393, D.D.C. (1975).

28. 396 F. Supp. 1393, 1396 D.D.C. (1975).

29. 396 F. Supp. 1393, 1399-1400 D.D.C. (1975).

30. 396 F. Supp. 1393, 1400 D.D.C. (1975).

31. 396 F. Supp. 1393, 1401 D.D.C. (1975).

32. Compare 39 Fed. Reg. 5706 (1974) with 40 Fed. Reg. 54183. Also, see Montgomery, Control of Agricultural Water Pollution: 545.

33. P.L.92-500 §208, 33 U.S.C. §1288.

34. Claudia Copeland, *Nonpoint Pollution and the Areawide Waste Treatment Management Programs under the Federal Water Pollution Control Act.* (Washington, US Government Printing Office, 1980).

35. Copeland, *Non-point Pollution*, pp. 23-25.

36. Beatrice H. Holmes, *Institutional Bases for Control of Nonpoint Source Pollution.* (Washington, DC, US Government Printing Office, 1979): 19.

37. P.L. 95-217 (1977).

38. P.L. 92-500 sec. 208(j)(1) as amended by P.L. 95-217 (1977).

39.   J. T. B. Tripp and A. Jaffe, "Preventing Groundwater Pollution: Towards a Coordinated Strategy to Protect Critical Recharge Zones." *Harvard Environ. Law Rev.* 3 (1979): 31, 34-35.

40.   James L. Conner II, "Using CERCLA to Clean Up Groundwater Contaminated Through the Normal Use of Pesticides." 15 *Environ. Law Rep. News & Analysis* 10100.

41.   By the Administrator of the EPA, among others. Conner, Using CERCLA to Clean Up Groundwater.

42.   *Protecting the Nation's Groundwater*, Vol. I, (1984): 11-13.

43.   *Groundwater Contamination in the United States*: 260.

44.   *Protecting the Nation's Groundwater*, Vol. I, (1984): 11.

45.   T. R. Henderson, J. Traubman, and T. Gallagher, *Groundwater: Strategies for State Action.* (Washington, DC: The Environmental Law Institute). (Hereafter referred to as ELI, Groundwater).

46.   ELI, Groundwater.

47.   Florida Administrative Rules §12-4.2545.

48   ELI, Groundwater: 112.

49   ELI, Groundwater.

50   N.Y.L. 1940 c. 727.

51   N.Y. Soil and Water Conservation District Law, section 9(7-a).

52.   N.Y. County Law section 223 (1982).

53.   N.Y. County Law section 299 (1982).

54.   N.Y. County Law section 224(8) (1982).

55.   *Protecting the Nation's Groundwater*, v. I, p. 107.

56.   Robert D. Hennigan, *Water Supply Source Protection Rules and Regulations Project: Final Report*, (Syracuse, NY State College of Environmental Science and Forestry, 1981): 22.

57.   Hennigan, Water Supply Source Protection, pp. 21-22.

58. Hennigan, Water Supply Source Protection, p. 48.

59. 6 NYCRR §703.5, 1978.

60. 6 NYCRR §703.5.

61. 6 NYCRR §703.11(a)2.

62. Assembly Office of Research, *The Leaching Fields: A Nonpoint Threat to Groundwater.* (Sacramento: California State Assembly, 1985): 11-35.

63. 1985 Cal. Stats. c. 1298, §1; Calif. Food and Agric. Code §§13143-13146 (West's Cum. Supp. 1988).

64. Calif. Dept. of Food and Agric. §13148 (West's Cum. Supps. 1988).

65. Calif. Food and Agric. Code §§13149-13152 (West's Cum. Supps. 1988).

66. California Health and Safety Code §§25249.5-25249.13, (West's Cum. Supp., 1988) added by Initiative Measure Proposition 65, 4 November 1986 (Prop. 65).

67. California Health and Safety Code §§25249.5 (West's Cum. Supp., 1988).

68. California Health and Safety Code §§25249.7(b) (West's Cum. Supp., 1988).

69. California Health and Safety Code §25249.5(d) (West's Cum. Supp., 1988).

Pesticide Regulation

PESTICIDES IN MODERN AGRICULTURE

Pesticides, or economic poisons, are used to control the infestations of insects, weeds, fungi and other undesirable forms of life which damage or compete with crops cultivated by man. Their use in agriculture helped increase yields, enhanced cosmetic quality and lowered food costs. Pesticides complement mechanization and help reduce labor used on farms.[1] The substitution of capital for labor in agriculture has been a result of monoculture agriculture, the continuous production of a single crop. Monoculture relies on uniformity of maturation and size, fields free of weed and insect damage, and mechanical harvesting. The use of insecticides, fungicides and herbicides have facilitated modern mechanized monoculture.

Modern agricultural techniques require increased pesticide usage for another reason. Agricultural practices most favorable to increased insects, weeds, pathogens, nematodes and other pest problems include monoculture agriculture, increased use of irrigation and inorganic fertilizer, multiple cropping, genetically uniform varieties, longer growing seasons, and reduced tillage.[2] Modern agriculture has replaced an ecosystem which was genetically, ecologically and economically diverse with one which is based on only a few species of plants and animals.

The increased intensity of agriculture has resulted in increased productivity, measured both per acre of land and per farm worker. While there has been progress in food production technology, the economics of running a farm business has also been changing. Agricultural enterprises have become more specialized. Food prices have dropped relative to other goods in the economy. The contribution of pesticides to lower food prices is difficult to estimate.

There are some differences between agricultural scientists as to the proper role of pesticide usage. Before the Second World War, farmers relied primarily on biological control for pest management. With the development of synthetic pesticides, the accepted method of controlling insect pests involved programmed spraying of insecticides at periodic intervals.[3] Reasons for this approach included prophylactic control, simplicity of decision making and the hope for eventual eradication of the target insect.

Programmed spraying can be contrasted with integrated pest management (IPM) techniques. IPM seeks to control pests at a certain level, known as the threshold. Other techniques, such as rotations, beneficial releases and selection of resistant varieties are used in addition to pesticides. Pesticide use is a last resort. IPM has slowly been gaining acceptance, but has faced numerous obstacles.

There is evidence that pest management behavior is not strictly profit maximizing, and farmers use pesticides to reduce risk.[4] This complicates predicting pesticide use. Farmers are subject to having their fears manipulated and influenced by advertising by chemical companies. Advertising appears to have stimulated aggregate demand in this oligopolistic industry.[5]

PESTICIDE EXTERNALITIES

While pesticides have apparently increased agricultural productivity, they have also created a number of social costs, or externalities. An externality can be defined as any benefit or cost imposed on one individual by another's action. The concept of externality is rooted in the utilitarian tradition that is the basis of neoclassical economics. As Baumol and Oates say,[6]

> An externality is present", when some individual's (say A's) *utility* or *production* relationships include real (that is, nonmonetary) variables, whose values are chosen by others (persons, corporations, governments) without particular attention to the effects on A's welfare. . . . The decision maker,

whose activity affects others' activity levels or enters their
production functions, does not receive (pay) in compensation
for this activity an amount equal in value to the resulting
(marginal) benefits or costs to others.

The concept has drawn criticism among environmentalists.
Neoclassical economics preoccupation with maximization is seen as
simplistic sophistry. Externalities can be used either to justify
pollution on efficiency grounds, a' la Coase, or to pay lip service to
critics who say economics fails to deal with values not reflected in the
market.[7] Whatever the shortcomings of externality theory, externalities
are frequently cited as a rationale for market intervention.

Calculation of the private benefits and costs of pesticide use is
relatively easy compared to estimating social cost. Market and
production data provide the information needed to estimate the value of
the marginal product, hence the benefits of pesticide use to the farmer.
Private cost is the amount of pesticide purchased times the price of that
pesticide. Innovations, specification error and imperfect information
introduce biases in the statistical derivation of private benefits. Despite
this, private benefits are more easily estimated than social costs and
benefits. Under competitive conditions market data give the private
optimal pesticide use.

The benefits and costs of pesticides can be quantified separately in
monetary terms and in terms of human health. Beneficial effects to
human health which are quantifiable would include a reduction in pest
borne diseases and injuries, decreased resources used to control diseases
and decreased debility. Detrimental effects to humans associated with
pesticide use include illnesses and poisoning from pesticides,
mutagenesis, carcinogenesis and teratogenesis. Such an analysis would
have to rely on epidemiological data, and subject to many of the
problems inherent in interpreting this kind of data. The lack of
observable human exposure and the need to interpolate from animal
studies is common to all toxic exposure studies. This is the best
possible given the ethical considerations that prohibit of intentional
exposure to humans under controlled circumstances.

The toxicological damage from pesticide use can be measured by
looking at the human exposure and toxicity of the pesticide. The
consumer must bear the increased health care costs caused by exposure
to pesticides. Exposure can occur in a number of ways: eating
contaminated food, by drift from aerial application, runoff into surface

water or leaching into groundwater. When they are included, the cost of damage to human health and environmental diversity are usually underestimated. Many question the ability to derive acceptable dollar figure for human health, premature death, or species extinction. These questions of value are at the heart of environmental economics and have sparked a long running debate.[8] At issue is more than just the balancing of public and private costs. Also questioned is the adequacy of the market as a measure of value.

The focus of most regulatory activity has been with food.[9] The exposure of consumers to toxic and oncogenic pesticides is the most immediate and far-reaching public health hazard of pesticides. It is by no means the only detrimental effect of pesticide use. The externalities created by pesticides are multi-faceted and complex, and go well beyond food consumption. However, pesticide use also threatens the health of farmers and farmworkers.[10] Pesticides can kill non-target species, or be accumulated in the tissue of animals higher in the food chain.[11] Chemical control technology can also disrupt the ecological balance in pursuing biological control several ways. For one, pesticides can kill beneficial insects. This often causes a resurgence of the pest that outstrips the replacement of beneficial insects. Farmers are forced to apply greater amounts of pesticides more frequently to prevent this resurgence.[12] This is frequently referred to as the "Pesticide Treadmill." The destruction of beneficials will also frequently cause a secondary outbreak of a pest formerly kept in check by its natural enemies.[13]

Pests also become resistant to the application of pesticides by genetic selection and mutation.[14] As pesticide resistance becomes more common, the cost of developing new pesticides increases to internalize their social costs, farmers can no longer rely on technology to resolve the problem of developing new biocides.

Regulations now control the distribution, sale and use of pesticides. This legislation, and subsequent administration, has attempted to reduce the externalities associated with pesticides.

The history of controversy surrounding pesticide regulation begins with the publication of *Silent Spring* by Rachel Carson.[15] When this bombshell exploded on the scene of science, it was received with praise and damnation, criticism and calumny, acclaim and esteem. While this is not the place to reproduce the replies and rejoinders to Ms. Carson's work, it's role in the passage of the current regulations of pesticides cannot be ignored. One congressman inserted parts of *Silent Spring* into the *Congressional Record*;[16] another congressman wrote a book designed to discredit Ms. Carson's work.[17] Her book was a major factor in the decisions to increase restrictions on pesticide use, ban certain chemicals such as DDT, dieldrin and aldrin, and be more cautious in approving new pesticides.

The decision to regulate pesticides more strictly was apparently not dominated by economic motivation. Other factors that led to that decision include a concern for future generations, wildlife and the environment that transcend economic thinking. However, economic reasoning was not absent from the legislative and regulatory decisions. Before looking at the existing regulations, let us turn to the economic rationale for regulating pesticide use.

REGULATORY DEVELOPMENT

Pesticide regulations were first developed to protect farmers from the sale of fraudulent concoctions by disreputable dealers.[18] As health and environmental problems associated with pesticide use mounted, the focus of pesticide regulation shifted from protecting the farmer to protecting consumers and the environment. Beginning with the Federal Insecticide, Fungicide and Rodenticide Act of 1947,[19] public health problems caused by pesticide use have been considered federal problems. FIFRA was drafted to not impose on pesticide manufacturers.[20] After its passage, pesticide laws were weakly enforced, with the expectation that they would be industry self-enforced. The Federal Committee on Pest Control was termed by its critics an "uncertain defender" of the environment.[21]

Environmental effects of pesticides came under federal regulation with the passage of the Federal Environmental Pest Control Act (FEPCA) of 1972.[22] FEPCA superceded many obsolete provisions of FIFRA. The act established a new agency to enforce pesticide regulations, the Office of Pesticide Programs in the US Environmental Protection Agency. The passage of FEPCA placed both existing and new regulatory functions in the hands of EPA.[23] Pesticide regulation has since had its share of interagency and intergovernmental conflicts. FEPCA explicitly preempts states rights to certain powers over pesticides, such as labelling and packaging requirements.[24] State regulations and enforcement programs may be stricter, but not laxer

than those at the federal level.[25] Central to the regulation of pesticides
is the registration procedure enacted by FEPCA. Registration requires
that the benefits and costs of a pesticide be considered. After a pesticide
is registered, the registration is routinely reviewed.

RISKS AND BENEFITS

The registration of a given pesticide is designed to ensure that the
benefits of using that pesticide outweigh the risks to human health and
the environment. Evidence is considered to decide whether or not to do a
risk-benefit analysis. Once risk-benefit analysis is triggered, the burden
of proof that benefits justify the risks is shifted to the manufacturer.[26]
Registration is lifted, and the manufacturer must collect data to rebut
the presumption of harm. This initiates a process known as Rebuttable
Presumption Against Registration, or RPAR.

Data required to register pesticides were laid out by the courts in
several cases brought by environmental groups within the first few
years after its enactment.[27] The EPA must decide on the registration of
all pesticides. The burden of proof for the safety of a pesticide rests on
the applicant for registration.[28] The court established guidelines for
environmental considerations required by an RPAR.[29]

RPAR attempts to resolve an environmental problem by removing
the registration of pesticides for which the costs and risks outweigh the
benefits.[30] Risk assessment in RPAR evaluates hazards to human
health, environment and structures, materials and crops presented by
using a particular pesticide. Risk regulation can be divided into risk
assessment and risk management. Risk assessment involves the
identification of the hazard, assessment of exposure to risk,
extrapolation to the population of toxicological or other risk data, and
the characterization of the risk presented by the activity. While full
information about any risk is virtually impossible, risk assessment can
be comprehensive or limited depending on the amount of information
available. Risk management seeks to alleviate risk based on the
technology available, costs, benefits, and political saliency.

The methodology practiced in pesticide regulation has more often
been labelled risk/benefit analysis.[31] This recognizes that, while the
benefits are reflected in market exchanges, the 'costs' are actually risks
borne by consumers and the environment.[32] Most economists

distinguish between risk and uncertainty. Risk is assumed to have a known probability distribution of events, and certain costs associated with each of those events. This information is unavailable under uncertainty. Environmental risk has been characterized by ignorance of the mechanism of adverse effects.[33]

It is not the purpose of this paper to explain,[34] critique,[35] nor defend risk-benefit analysis. It is instead intended to offer a framework to understand the uses and abuses of such analysis. While economists are quick to recognize the methodology's shortcomings, few advocate its abandon. Economists disagree over how to perform risk-benefit analysis under conditions of uncertainty.

The benefits of pesticides are assumed to be measured by their productivity compared with if they are removed from the market. Pesticide productivity pesticides has been estimated from experimental data and from market data. Pesticide productivity data from experiments are subject to numerous problems. Among these are the differences in experimental and commercial management practices and the lack of data on economically relevant changes in product quality in experiments. Despite these problems, experimental data are often the only source that permits agricultural economists to do *ex ante* research on pesticides.

Productivity statistics from market data also have their disadvantages. The data are frequently aggregated, and much variation between products is lost. The models require strict assumptions be made about the homogeneity of technology, products, tastes and behavior. These problems have not prevented agricultural economists from estimating the short-run[36] and long-run[37] productivity of pesticides and insecticides from market data.

Risk estimation focuses on human health hazard. The estimation of the hazards must be made on a case by case basis. No two molecules will act exactly alike. Nevertheless, the possession of certain closely related chemical, physical and biological properties which pesticides have in common make comparative data of great value in prediction.[38] Data are required for the health and environmental effects of pesticides. Health effects analyzed include mammalian LD_{50}, carcinogenesis, teratogenesis and mutagenesis. Environmental effects analyzed include soil metabolism, soil resistance, degradation in water, and photodegradation.[39] Data collection on the health and environmental effects of pesticides is an expensive and time consuming process that firms wish to avoid. The RPAR process has been accused of discouraging product since its enactment in FEPCA. Moreover, FEPCA has failed in many cases to limit consumer exposure to pesticides in food.[40] The registration process also fails to take water contamination into account.

The explicit economic considerations included in FEPCA set a precedent for environmental policy. Never before had benefit-cost analysis been required for regulation. Prior to that, benefit-cost analysis was used as criteria for public investments. While some of these projects considered environmental amenities and value of human life, pesticide registration required that new methodologies be developed for estimating costs to human health, wildlife, and environmental quality. These costs are not priced in any market, and therefore must be estimated by proxy. Many of these attributes cannot be quantified. The value of environmental degradation will not be adequately estimated in benefit/cost analysis.[41]

Toxicologists extrapolate adverse health effects from animal bioassays. The data from these bioassays may bear little resemblance to the health effects in humans.[42] Policy analysts must operate under scientific uncertainty over the level of risk and magnitude of effects in formulating health and safety regulations. The economic analyses must be made *ex ante*. Not only the risks, but even the benefits of pesticides are uncertain.[43] Pesticide regulation are based on this incomplete information. The application of risk-benefit analysis has been labelled a 'random guessing game' as a result of this discrepancy.[44]

One major source of disagreement in performing risk-benefit analyses has been on the value of human life. There are wide disparities in values derived by economists and used by regulatory agencies to implicitly or explicitly value human life.[45] Economists generally prefer to eschew direct money-for-life exchanges, and phrase their results in such terms as 'value of risk avoided.'[46]

Another problematic issue is the estimation of option prices, or the *ex ante* willingness to pay for the preservation of an unreplaceable resource. Contingent valuation is the most widely used technique for these estimations. Simply put, people are asked in surveys how much they would be willing to pay for an environmental amenity, or what they would have to be paid to replace it. Hypothetical demand curves are derived from these surveys, and are used to estimate benefits of environmental regulation. Actual interpretation of this method has been subject to question repeatedly by economists.[47] Despite the obvious flaw of suppositious results, and the arcane flaw of dubious interpretations, few economists propose an abandonment of contingent valuation. Some half-hearted defenses appear as damning with faint praise.[48]

Cost-benefit analysis can organize available information. Cost-benefit analysis cannot overcome uncertainty, but it can make uncertainty more manageable.[49] Without data, however, the usefulness of cost-benefit analysis is limited. Because data collection on the extent to which pesticides are in groundwater is very preliminary, there is no accurate estimate of exposure. Without exposure data, it is impossible to perform the necessary risk assessment. Regulators may actually be better off with qualitative risk assessment in the absence of hard data.[50]

Regulatory agencies must depend on the private sector for toxicity data. After the Industrial Bio-Test (IBT) case, the credibility of such data cannot be taken for granted. IBT was an independent laboratory that performed toxicological studies under contract for several major chemical companies. Investigations revealed that IBT routinely falsified data and subsequently produced test results favorable for the companies' market positions. IBT employees held responsible for the falsification were convicted on criminal charges. Environmental groups sued to remove the registrations of pesticides approved as a result of IBT studies.[51] However, the EPA has lacked critical technical expertise to audit results.[52] This observation was made prior to the Reagan revolution in regulatory policy. Subsequent hearings before Congress have seen environmental groups asking for more oversight, corporations say revised protocols provide adequate safeguards, and the agency reluctant to admit it unable to do its job, but pointing out that its resources are not sufficient to achieve its statutory mandate.[53] Corporate control of the regulatory agenda, and dependency upon corporate technical expertise have, if anything, increased since then. Environmental groups cannot compete with the in-house technical expertise of the regulated industry.

Groundwater contamination was not mentioned as a consideration for much risk assessment literature when RPAR was first implemented. Human health risk focused on food contamination and worker exposure. One government official seemed to suggest that 'involuntary exposure' was more likely an emotional fear of the public that rested on few expert's facts.[54] Where water was mentioned at all, the greater risk was assumed to be runoff into surface water. This more often was used in conjunction with risk to wildlife than drinking water contamination.

The pesticide registration process is not applicable to the use of an approved pesticide in an approved manner which nevertheless pollutes water. The procedure to screen out pesticides deemed too dangerous for the public good has not been fully implemented yet. Even after it is implemented, not all pesticides that pose health hazards will be prohibited. FEPCA does not specifically address the setting of standards for residues in any medium other than food. Pesticide residues in food have lower tolerance levels than those in the environment. Given limited resources to enforce controls in food or in water, food usually has priority. The toxic effects of pesticides may be reduced, but not eliminated by FEPCA. Pesticide residues in water will remain in spite of the safeguards undertaken by FEPCA.

The RPAR process is dynamic, while risk/benefit analysis relies on static models. Static analysis cannot take into account the technological change that makes *ceteris paribus* assumptions obsolete. The RPAR process is costly and drawn out. It is designed to protect the rights of manufacturers and, in doing so, may take away from the rights of the public. Risks and benefits are subject to great change over the review period.[55] Because of the methodological difficulties of estimating risks of pesticide use and the relatively available estimates of pesticide benefits, benefits were likely to be overestimated and risks underestimated in the initial registration and RPAR process. In particular, groundwater contamination costs and the risk of exposure appeared outside of the scope of the registration and RPAR processes.

Drinking water protection falls into a separate category of regulation. The Safe Drinking Water Act, (SDWA), not FEPCA, is responsible for the setting of standards for water. Comparing the regulations promulgated by the two acts, pesticide residues for food have lower limits than those in the environment. Given limited resources to enforce controls in food or water, food usually is given priority. Pesticide residues will be found in water as long as pesticides are used. The acceptable amount of pesticides in drinking water is a matter of politics, technology, economics and culture.

NOTES
CHAPTER III

1. J. C. Headley and J. N. Lewis, *The Pesticide Problem: An Economic Approach.* (Baltimore, Johns Hopkins University Press and Resources for the Future, 1966): 75.

2. A. W. A. Brown, *et al.* "Plant Protection from Pests" in A. W. A. Brown (ed.) *Crop Productivity -- Research Imperatives.* (East Lansing, MI and Yellow Springs OH, Michigan Agriculture Experiment Station and the Charles F. Kettering Foundation, 1975).

3. John Perkins, *Insects, Experts and the Insecticide Crisis,* (New York: Plenum Press, 1982).

4. A review of the literature of risk reduction in pesticide use can be found in Gerald A. Carlson's article "Risk Reducing Inputs Related to Agricultural Pests" in *Risk Analysis for Agricultural Production Firms: Concepts, Information Requirements and Policy Issues.* (UrbanaChampaign: University of Illinois Dept. of Agr. Econ., 1984): 164-176.

5. Robert D. Boynton and Elizabeth A. Schiferl, *The Incidence of Rivalry in Pesticide Advertising.* Cornell University, Department of Agricultural Economics Staff Paper 83-14 (Ithaca, NY, Cornell University, 1983).

6. William J. Baumol and Wallace E. Oates, *The Theory of Environmental Policy.* (Englewood Cliffs, NJ, Prentice-Hall Inc., 1975): 17-18. Italics the authors'.

7. A critic of the application of externality theory to pesticide use has been Roland Clement of the Audobon Society. His review of *The Pesticide Problem* appeared in *The Natural Resources Journal,* 8(1): 11-22. A critic of externality theory and neoclassical theory's approach to pollution and resource problems is Hazel Henderson. See *Creating Alternative Futures,* (New York, G. P. Putnam, 1978).

8. This debate is briefly summarized in Brian Baker, "Risk, Uncertainty and Technology Assessment," *Environment, Technology and Society.* 52: 8-10.

9. National Research Council, Board on Agriculture, *Regulating Pesticides in Food.* (Washington: National Academy Press, 1987).

10. Molly Coye, "The Health Effects of Agricultural Production," in Kenneth Dahlberg, *New Directions for Agriculture and Agricultural Research* (Totawa, NJ: Rowman and Allenheld, 1986).

11. Papendick, Elliot and Dahlgren, "Environmental Consequences of Modern Production Agriculture: How Can Alternative Agriculture Address These Concerns?" *American Journal of Alternative Agriculture* 1: 3. Carson, *Silent Spring*.

12. Paul DeBach, *Biological Control by Natural Enemies*. (New York: Cambridge University Press, 1974).

13. Mary Lou Flint and Robert van den Bosch, *Introduction to Integrated Pest Management*. (New York: Plenum Press, 1981).

14. George P. Georghiu, "The Magnitude of the Resistance Problem," in *Pesticide Resistance: Strategies and Tactics for Management of Pesticide Resistant Pest Populations*. (Washington, DC: National Academy Press, 1986).

15. Boston, Houghton Mifflin and Co., 1962.

16. John V. Lindsay, quoted in Frank Graham, *Since Silent Spring* (Boston, Houghton Mifflin and Co., 1970): p. 50.

17. Jamie L. Whitten, *That We May Live* (Princeton, D. van Nostrand Company, 1966).

18. For a summary of the history of pesticide regulation, see National Research Council, *Regulating Pesticides* (Washington, US Government Printing Office, 1980):20-28; Jeffrey D. Huffaker, "Regulation of Pesticide Use in California", 11 *UC Davis Law Rev.* 273, 275-277 (1978).

19. Current version at 7 U.S.C. §136.

20. Christopher Bosso, *Pesticides and Politics: The Life Cycle of a Public Issue*. (Pittsburgh: University of Pittsburgh Press, 1987): 45-60.

21. Graham, *Since Silent Spring*: 185-191.

22. 7 U.S.C. §§ 136-136y. (1976).

23. Huffaker notes that FEPCA is better suited to restrict the sale of adulterated poisons than to prevent the use of dangerous poisons. Regulation of Pesticide Use in California, 277.

24.   7 U.S.C. 136v (b).

25.   7 U.S.C. §136v(a)

26.   Environmental Defense Fund v. Ruckleshaus (1971) 439 F 2d 584 (DC Cir.)

27.   *Environmental Defense Fund v. Ruckelshaus*, 439 F. 2d 584 (D.C. Cir., 1971); *Environmental Defense Fund v. US Environmental Protection Agency*, 510 F. 2d 1292 (D.C. Cir., 1975); *Environmental Defense Fund v. US Environmental Protection Agency*, 548 F. 2d 998 (D. C. Cir., 1976).

28.   *Environmental Defense Fund v. Ruckleshaus*, 439 F2d. 584.

29.   *Environmental Defense Fund v. EPA*. 510 F2d. 1292 (D.C. Cir. 1975).

30.   National Research Council, *Regulating Pesticides:* 23.

31.   Risk/benefit analysis is an outgrowth of cost/benefit analysis. Most recent texts which cover cost/benefit analysis at least outline risk/benefit analysis. See Daniel Swartzman, Richard A. Liroff and Kevin G. Croke, *Cost Benefit Analysis and Environmental Regulations: Politics, Ethics and Methods*, (Washington, DC: Conservation Foundation, 1982); A. Myrick Freeman, "Risk Evaluation in Environmental Regulation, in Wesley A. Magat (ed.) *Reform of Environmental Regulation*, (Cambridge, MA: Ballinger Publishing, 1982); Talbot Page, "A Generic View of Toxic Chemicals and Similar Risks", *Ecology Law Quarterly* 7: 207-244 (1978).

32.   National Research Council, *Regulating Pesticides*.

33.   R. Talbot Page, "A Generic View of Toxic Chemicals and Similar Risks," *Ecology Law Quarterly* 7(1978): 207-244.

34.   Many texts are available which explain the methodology. See, for example, Edward M. Gramlich, *Benefit-Cost Analysis of Government Programs*. (Englewood Cliffs, NJ: Prentice Hall, 1981); Ezra Mishan, *Cost-Benefit Analysis*. (New York: Praeger, 1976); Edith Stokey and Richard Zeckhauser, *A Primer for Policy Analysis*. (New York: W. W. Norton, 1978). For specific applications to environmental policies, see A. Myrick Freeman III, *The Benefits of Environmental Improvements*. (Baltimore: Johns Hopkins Press, 1979) and Allen V. Kneese, *Measuring the*

Benefits of Clean Air and Water. (Baltimore: Johns Hopkins Press, 1984).

35.   A cogent critique of cost-benefit analysis can be found in Michael S. Baram, "Cost-Benefit Analysis: An Inadequate Basis for Health, Safety, and Environmental Decisionmaking." *Ecology Law Quarterly* 8: 473-531 (1980) and in Stephen J. Kelman, "Cost-Benefit Analysis: An Ethical Critique" *Regulation* 36 (1981).

36.   J. C. Headley, "Estimating the Productivity of Agricultural Pesticides," *Am. J. Ag. Econ.* 50(1) (February, 1968): 13-23.

37.   G. A. Carlson, "The Long-Run Productivity of Insecticides," *American Journal of Agricultural Economics* 59(3) (August 1977): 543-548.

38.   Eugene E.Kenaga, "Evaluation of the Hazard of Pesticide Residues in the Environment" in Watson and Brown (eds.), *Pesticide Management and Insecticide Resistance* (New York, Academic Press, 1977): 90.

39.   40 CFR §162.11 et. seq. spells out the requirements of RPAR.

40.   National Research Council Board on Agriculture, *The Delaney Paradox.*

41.   See Michael Baram, "Cost-Benefit Analysis: An Inadequate Basis for Health, Safety, and Environmental Decisionmaking." *Ecology Law Quarterly* 8 (1980): 473-531.

42.   Robert A. Neal, "Risk Benefit Analysis: Role of Regulation in Pesticide Registration," in S. Kris Bandel, Gino J. Marco, Leon Golberg and Marguerite L. Leng (eds.) *The Pesticide Chemist and Modern Toxicology,* (Washington, American Chemical Society Symposium 160, 1981): 471. The author goes on to say that quantitative risk assessment is "a useful exercise" despite this and other limitations.

43.   Erik Lichtenberg and David Zilberman, "Problems of Pesticide Regulation: Health and Environment versus Food and Fiber," in Tim T. Phipps, Pierre R. Crosson and Kent A. Price (eds.) *Agriculture and the Environment.* (Washington, Resources for the Future, 1986): 123-145.

44.   Liebe Cavelieri, *The Double-Edged Helix.* (New York, Columbia University Press, 1981).

45. Three summaries of the literature on the value of human life give the fundamental elements of the models used: Joanne Linnerooth, "The Value of Human Life: A Review of the Models", *Economic Inquiry* 17: 52-74 (1979); W. B. Arthur, "The Economics of Risks to Life", *American Economic Review* 71(1): 54-64 (1981); John D. Graham and James W. Vaupel, "Value of a Life: What Difference Does it Make?" *Risk Analysis* 1(1): 89-95 (1981).

46. Erik Lichtenberg and David Zilberman, Efficient Regulation of Environmental Health Risks. (San Francisco, Western Consortium for the Health Professions, 1985).

47. Most recently by Jean-Paul Chavas, Rich Bishop and Kathy Segerson,"*Ex Ante* Consumer Welfare Evaluation in Cost-Benefit Analysis," *Journal of Environmental Economics and Management* 12: 255-268 (1986); and V. Kerry Smith, "Uncertainty, Benefit Cost Analysis and the Treatment of an Option Value," *Journal of Environmental Economics and Management* 14: 283-292 (1987).

48. See Jon Conrad, "On the Evaluation of Government Programs to Reduce Environmental Risk." *American Journal of Agricultural Economics* 68: 1272-1275 (1986).

49. U.S. General Accounting Office, *Cost-Benefit Analysis Can Be Useful in Assessing Environmental Regulations, Despite Limitations.* (Washington, DC: U.S. Government Printing Office, 1984): 11-12.

50. Mary O'Brien, "If Not Risk Assessment, then What?" *Journal of Pesticide Reform* 10 (1990): 2-6.

51. See Eliot Marshall, "The Murky World of Toxicity Testing." *Science* 220: 1130-1131 (10 June 1983); Keith Schneider, "The Case Against Industrial Bio-Test." *Amicus Journal* 4(4): 14-26 (1983).

52. National Research Council, *Decision Making in the Environmental Protection Agency.* Washington, DC: National Academy of Sciences: 51-57 (1977).

53. U.S. Congress. House. Committee on Agriculture. Subcommittee on Department Operations, Research and Foreign Agriculture. *Regulation of Pesticides* v. IV (Washington, DC: U.S. Government Printing Office, 1983): 413-616.

54.   Stanford Fertig of USDA, "The RPAR Process and the Impact of Regulatory Decisions on Agriculture," *Proceedings of the Southern Weed Science Society* 34: 9 (1978).

55.   See, for example, Linda Lorraine May, *Dynamic Risk Benefit Analysis in Pesticide Regulation: The Case of Toxaphene.* (River,side, University of California Unpublished Ph.D. Dissertation, 1984).

Drinking Water Standard Setting

Standards have developed into regulatory tools for health, safety, and consumer protection. Seemingly small differences in a standard can make great differences in both the costs and benefits of a regulation, however measured. The level of such standards has become an issue of concern for firms and consumers, industry and citizen's groups, legislative bodies and administrative agencies. Environmental standard setting has a large cast of actors, with widely varying perspectives of how standards are actually set.[1] The role of analysis can be explicitly stated, yet the results of analysis can be applied differently.

SAFE DRINKING WATER ACT

The Safe Drinking Water Act (SDWA) illustrates the tension between what information is available and by whose values that information is interpreted. Federal regulations to protect drinking water are contained in the SDWA.[2] The Environmental Protection Agency (EPA) is the agency responsible for its administration. With a few exceptions, the SDWA does not give EPA explicit regulatory power to prevent drinking water contamination. Because society has limited resources to spend on clean water, a zero contamination policy is not feasible. The SDWA stated purpose is to "protect health to the extent feasible, using technology, treatment techniques, and other means which the Administrator [of EPA] determines are generally available (taking costs into consideration) . . . "[3]

While the SDWA is set up to encourage a scientific and rational basis for standard setting, the actual process has been marked by uncertainty and dissensus. It is a regulatory law designed to alleviate risk to public health. As an act that articulates broad policy goals, it has been marked by consensus. As a regulatory policy, it has generated much debate. The SDWA has sought to implement standards through objective, scientific information. In practice, information has not been sufficient to promulgate standards for most chemicals found in drinking water.

The basic mechanism by which EPA sets drinking water quality standards is the Maximum Contamination Level (MCL) for various contaminants.[4] The EPA sets MCLs for all public water systems. A scientific advisory board was created to establish a uniform approach for setting standards. This board makes Recommended Maximum Contaminant Levels (RMCLs) based on scientific evidence.[5] These are guidelines that do not have the force of law, but are intended to provide the water supply operator with a rule of thumb for water quality. RMCLs are more flexible than MCLs, but lack legal sanctions.

The methodology used by the advisory board is known as Acceptable Daily Intake (ADI).[6] This involves extrapolation of the No Observable Adverse Effect Level to the population using exposure data to set the Acceptable Daily Intake. The Acceptable Daily Intake is then adjusted by an uncertainty factor. At every step of the way, there are several choices must be made, e.g. over which toxicological study, which exposure level, which safety factor should be used.
Attempts to build consensus where data are remarkably inconsistent have been notably difficult.[7]

MCLs are supposed to be set as close to RMCLs as is technically feasible, considering cost. RMCLs become interim standards during a period of public review and comment. To make what was recommended mandatory, EPA was given the additional burden of analyzing technology, cost and implicit benefits. If there is already disagreement over the level at which an RMCL should be set, disagreement over technical feasibility and cost further muddy the waters.

CONCEPTUAL FRAMEWORK

Standards are based both on information available to the decision-maker and the degree of consensus or conflict between parties affected by the level of the standard. This has ramifications on both the utility and validity of the models used to set standards. Formal cost-benefit analysis performs different roles given different levels of information and consensus. Conflict over values can lead to a policy stalemate broken by the rejection of quantitative risk assessment.

The United States is unique among developed countries in the degree of analysis and weight of economics in the regulation of toxic chemicals.[8] This is as much if not more a result of American legal and political traditions than a result of the scientific and economic nature of toxic chemicals management.[9] While administrators must weigh economic and technical considerations, they are left wide latitude in the final regulations. Instead of strengthening discretion, such vagueness leaves regulations open to frequent legal and political challenges.[10] Rather than facilitating decision-making, economic analyses have been used, some would say misused, to delay or obstruct regulations.

If the parties involved in risk management process share common values, and a consensus on priorities emerges, then risk management may simply require the implementation of available technology. If participants conflict over values and priorities, then risk management will be inherently political.

To withstand court challenge and judicial review, regulation cannot be 'arbitrary' or 'capricious.'[11] Regulation is often intended to decrease the likelihood of risk. The cost to the regulated increases with the degree of regulation. Some risks may be so great as to merit that their associated activities be prohibited. However, closing down an activity is neither a realistic nor desirable solution in many situations. Standards permit risky activities while holding risk to acceptable levels.[12] As regulations become more complicated and costly, there is a greater need for cooperation between lawyers, scientists, engineers and economists. More important, perhaps, is bridging the gap between experts and citizens. While standards and guidelines are set to protect the public within levels of acceptable risk, the very act of setting standards changes the public's notion of acceptable risk.

Typology of Standard Setting
Risk Management
Conflict
Consensus
Risk Assessment
Certain
Uncertain
Pluralistic Bargaining
Technocratic Rationality
Speculative Augmentation
Bounded Rationality

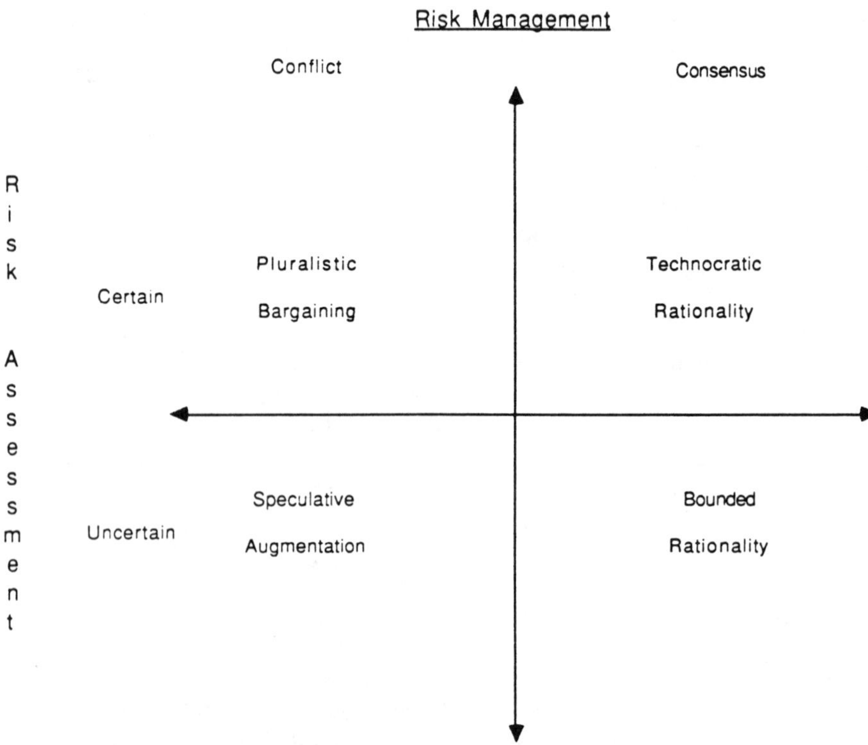

Figure 4.1

Models to set standards include technocratic rationality, bounded rationality, pluralistic bargaining, and speculative augmentation (figure 4.1). These models should be familiar to any student of policy analysis. Each by itself fails to consistently explain the role of analysis and agreement in rule-making. The typology synthesizes these divergent models of decisionmaking that depend on processing information and consensus. While one model may 'fit' regulatory action at one time, political conditions may change to make that model less apt to describe action.

TECHNOCRATIC RATIONALITY

When standards are based on scientific certainty made with a consensus over values, policy is left to technocratic experts. Technocratic rationality is the setting of standards to regulate risk at a socially optimum level. This model draws heavily on theories of welfare economics discussed in chapter 3. Market intervention can be justified as economically efficient in a neoclassical sense when social costs outweigh the social benefit of permitting an activity to continue unabated.

The economic theory of externalities implicitly assumes that there is an optimal level of pollution. Market failure prevents the parties from bargaining and compensation that would correct the externality efficiently.[13] Models to optimize social welfare or aggregate utility have a certain intellectual appeal for their rigor and elegance. Policy objectives are represented by a single easily maximized function subject to clearly defined constraints. The model assumes that the regulatory agency has at its disposal perfect information. Society's preferences are defined by a well-behaved, continuously differentiable social welfare or aggregate utility function that can be represented by a single value and maximized. The optimal level of regulation is solved by equating the marginal utility of income with the marginal disutility of risk. This solves for the maximum net benefits of a policy.

Pareto optimality, where nobody can be made better off without someone else being made worse off, implies a consensus. The market, under conditions of perfect competition without externalities will achieve a Pareto efficient allocation.[14] Unanimity rules are a means of efficiently reallocating property outside the market.[15] Scientific knowledge, in the form of a consensus of neutral, informed experts, forms the basis for technocratic rational policy.[16] Rational government intervention does not interfere with the market economy, but instead maximizes social welfare and efficiently corrects market failure. Benefits from economic activity may so far outweigh risk associated with a given activity that any interference with the market would result in a loss of welfare to society. At the other extreme, hazard from certain activities may be so great as to preclude those activities from being pursued. Regulation acknowledges that the market provides a suboptimal allocation of resources, and that total prohibition is also suboptimal. The decision of whether or not to regulate, and, if so, how much is appropriate is then a matter of knowing society's preferences and regulation's impact on these preferences.

The SDWA was created as a scientifically based program that would efficiently achieve optimal water quality. It did not mandate a zero level of risk, unlike the Delaney clause for food additives. Instead, it left to technocrats in government the discretion to rationally set standards. The legislature directed that MCLs stress the protection of health, but also recognize the consideration of costs.[17] The courts interpreted the SDWA as giving EPA the discretion to set standards 'efficiently.'[18] Congress also intended that drinking water contamination levels be set by civil servants, not left to the market. The Drinking Water Advisory Council was established as a body of technical advisors.[19] Debate about the Advisory Council's functions showed a faith in expert solutions.[20] While the legislation and decisions may not explicitly use the argot of welfare economics, they promote technocratic rationality. This is consistent with the overall view of the functioning of EPA.[21]

Technocratic rationality suffers from several practical problems. The model may serve as a valuable normative tool for policymakers, but the lack of consensus and information are usually too great for standards to be set based on a social optimum. No Federal programs that regulates toxic substances has achieved the full consensus needed for rational decision-making. Methodologies to rationalize standard setting have evolved under circumstances that are, if anything, more polarized.[22] Not surprisingly, the EPA faced critics in its implementation of the SDWA. Some felt the agency lacked scientific evidence to set any standards.[23]

BOUNDED RATIONALITY

Risk assessment has not readily conformed to quantification. It is impossible for government to obtain and process all information on risks, costs, and benefits because of limited resources. Expected utility models fail to account for the cost to gather and analyze data to select the optimal policy. Data requirements for setting environmental standards are great. A regulatory agency seeks to meet some short-term goal, rather than maximize some social welfare function. An agency with limited resources will seek to define a small number of easy to examine policy options and rank them by some criteria. After receiving feedback from the regulated industries and the beneficiaries of the regulatory policy, the agency adjusts the policy to make up for perceived shortcomings. Policymakers will often discuss proposed standards well before they are formally promulgated. Uncertainty is avoided. As one analyst put it, "surprises can be expensive."[24] This approach is an adaptation of bounded rationality.[25]

Bounded rationality involves the setting of standards to meet some set goal. Complete information is not desirable if collection is expensive, time consuming, and leads to long delays. Bounded rationality begins by defining the problem in manageable terms, and permits decision-making with limited information.[26] Bounded rationality can be thought of as making marginal changes in setting standards that correct for social costs. Efficiency is only one of several criteria that policy makers use. It may not be the highest priority. Rigidity of both the organizations that set standards and the organizations that must conform to them makes bounded rationality a better model for administrative behavior.

The criterion of efficiency cannot be used for many types of decisions.[27] Bounded rationality features satisficing between two conflicting goals to reach a greater efficiency. While bounded rationality does not deny the use of optimization in decision-making, it recognizes disequilibrium conditions in the enactment of policy, puts a great deal of emphasis on individual interest and incentive, and has made economists more aware of the value of information in decision making. Refutation of both technocratic rationality and bounded rationality in certain cases is no reason to abandon the search for improved public information. Citizens delegate decisions to experts in many areas and expect them to know better than the general electorate. This delegation of authority causes democratic society to select values, but requires limits on those making the decisions. Normative models have a place, both as a guide for reaching preconceived goals, and to help understand behavior guided by these models.

The SDWA, the EPA lacked adequate information to set further standards after the law was initially passed. The agency still had a broad consensus to set standards, but limited resources prevented data collection and interpretation to set standards at an optimal level. Despite this, EPA set standards for trihalomethanes in 1979.[28] The standards set for trihalomethanes are notable not just for being the only standards set after 1975; they also carry with them a greater literature of analysis than other contaminants. Risk assessment for toxic chemicals is dynamic because both the ability to measure contamination and predict chronic toxicity have consistently improved.[29] Despite these advances, the actual risk posed remained speculative at the time the rules were made.[30] Given the constant improvement in scientific capability, it is tempting for administrators to wait for 'better' data to come along. Cost-benefit analysis of the new rules relied on widely varying estimates of health risk.[31] Scientific uncertainty was obfuscated by economic uncertainty about intertemporal equity and the value of a life saved.[32] Regulators responsible for deriving the standards acknowledged the usefulness of cost-benefit analysis, but were aware of its limitations.[33]

The Environmental Protection Agency faced a reduction in budget and staff. Although inaction was within the authority of the agency's discretion, this had not been the intention of the legislature. A policy stalemate stalled the setting of MCLs. If RMCLs represented 'rational' standards, MCLs were not set based upon RMCLs because either the technology to meet those levels did not exist, or the cost of the available technology was prohibitive. This explanation implicitly assumes that all sides agree on technical feasibility, costs, and perhaps most importantly, benefits. Congressional testimony shows this was not the case. Regulators were frequently put in the position of 'putting out fires' by dealing with the 'chemical of the week.' Many mandated tasks have not been carried out by the agency.[34] Analyses submitted by

EPA have lacked a firm methodology and consistency in promulgating rules. Much of this stemmed from a lack of time and adequate resources to do a systematic analysis. Collection of data on risk has developed into more of an analytic culture than a methodology.[35]

While more pragmatic in its methodology than technocratic rationality, bounded rationality still suffers from many similar weaknesses. The goal orientation of bounded rationality requires that explicit criteria be drawn. This is bounded rationality's answer to the Social Welfare Function. Distributional considerations, environmental amenity preservation, and health and safety protection may not reflect cultural norms or universally held values. As with the social welfare function, normative models used by bounded rationality require consensus of values to guide decisions.

The lack of consensus over goals requires political bargaining. This model of regulation does not require the maximization of any single value or set of values in setting standards. Rule-making procedure in this model is more important than the outcome. Political and economic power of different interest groups involved in the regulatory process determines policy enactment. Regulations are proposed, discussed, reviewed, revised and promulgated with contending parties seeking to influence the final outcome.

PLURALISTIC BARGAINING

Standards can be set to balance the interests of two or more conflicting factions. This model of regulation has evolved from two earlier theories: the public interest theory and the capture theory. In the public interest model, the regulatory agency enacts laws to advance the general welfare of all society, over and above benefits to its individual members. In the capture theory, regulatory agencies are prisoners of the industries and firms they regulate. They act to maximize regulated firms' profits. An interest group's bargaining power depends on the resources it has available. Access to the policy making structure determines the salience of an interest group's position. The bargaining process does not automatically favor either the general public, any organized interest seeking regulation, or the regulated industry.[36] This model has its roots in the concepts of countervailing power[37] and interest group liberalism.[38]

According to this model, costs and benefits to society are less important to the concerned interest groups than how those costs and benefits are distributed. Movement from consensus to dissensus is caused by the redistributional consequences of regulation.[39] Technocratic rationality assumes a positive sum game where losers, if there are any, can be compensated by winners. Pluralistic bargaining is a zero-sum game where dissensus is driven by potential gains and losses.[40] The economic theory of politics can explain distribution effects with enough economic interest variables in the model. Such an explanation begs the question of political action beyond narrow self-interest.[41] Groups will pressure the agency responsible for setting standards based upon individual gains and losses. Under pluralistic bargaining, the agency seeks to delay implementation until data are as complete as possible.

Risk assessment in pluralistic bargaining is used to both determine the distributional consequences of risk management, and to delay regulation under the presumption of complete knowledge. Political power and bargaining strength determine the degree and manner of regulation. The regulatory agency seeks power by political arbitrage. This requires legitimation to carry out its mandated responsibilities. For broadest support, it must position itself between politically adverse groups, unaligned to any player, brokering disputes.

As with the capture theory model, the regulated industry has several advantages over other actors. Direct capture of the regulatory apparatus is too crude for corporations. Industry groups usually have few members and are well organized, making them better equipped to lobby government.[42] Regulated firms seek to co-opt academic experts.[43] Corporations can frequently blackmail government agencies by threats to use shear economic size and market power.[44] The regulated industry also possesses an asymmetrically large portion of the data and technical expertise needed to interpret the data.

The uneven bargaining power in balancing these tradeoffs between industrial and environmental interests are manifested in toxic substances risk assessment. The premise supporting safe minimum standards of exposure is the statistical relationship between exposure and mortality. This requires vast amounts of data. Regulatory agencies must depend on the private sector for toxicity data.

Corporations have a clear stake to delay implementation of standards.[45] Industry representatives of industrial interests are reluctant to represent a position that might be construed as favoring contaminated drinking water. To argue for standards based on an objective, rational, scientific basis is to avoid hastily set standards that may result in unnecessarily high costs.[46] Corporations have uneven bargaining power in the process of setting standards. Many are multinational and can threaten to move plants and jobs abroad if standards are too strictly set. Corporations have greater financial and scientific resources to draw upon when MCLs are being considered. Business can use these resources for public relations, lawyers, political consultants and lobbyists, as well as scientific expertise.

These advantages do not inevitably lead to decay and capture. Corporations express a preference for market, rather than governmental regulation.[47] The very existence of regulatory mechanism is evidence that corporations don't control the regulatory agenda. Clear policy directives and reliance on the rule of law can prevent regulatory authority from being co-opted by the regulated.[48] The political support of a well-organized constituency can successfully compete with the regulated industry provided it can participate in the regulatory forum in a coherent, informed manner. The regulatory agency is left in a position to balance competing clientele interests.[49]

The movement from consensus to dissensus caused a long period where standards were not set. When the distribution of the costs of new standards became clear to industry and local governments, these interest groups worked to block any new standards. Review, comment and oversight leaves administrative agencies the final say in setting standards. The system is by no means closed, however. All standards are set in an environment of bargaining and negotiations that are open and legal, albeit esoteric and technical. Bargaining over environmental standards involve batteries of scientists, reams of data and long debates over the interpretation of the results of experiments and tests. A regulatory agency charged with setting environmental standards faces the task of collecting massive amounts of data, assembling and analyzing a vast array of information while it mediates conflicting interests. As a regulatory agency matures, the amount of data it has at its disposal increases. More than data are required; a political consensus must be present for standards to be derived from the data. Standards are based on political strength rather than scientific evidence where there is dissensus. More information makes the conflicted parties more sensitive to their gains and losses. After EPA promulgates standards for drinking water quality, the burden is then on the individual supplier to meet these standards. EPA received a strong mandate for implementing standards; but the first round of standard setting in 1975 showed it was uncomfortable implementing a broadly defined law.

Fiscal conservatives opposed SDWA for imposing Federal rules on what is perceived by many to be a local problem, and for costing the Federal Government too much in the subsidy of costly upgrades of water systems. Another source of opposition to MCLs have been the local governments responsible for meeting the standards. As water suppliers, they bear much of the cost of MCLs, and are directly accountable to the public for water contamination. Where drinking water quality has become a political issue, incumbents have been vulnerable to attack for failure to protect public health and safety. Suppliers are often put in the embarrassing position of having to defend supplying water that exceeds an RMCL.[50] The system complies with the law; but it is contaminated beyond a recommended level, therefore it might be unsafe. Adequate data to analyze a salient position were unavailable. While supporting the MCLs in principle, local governments have long criticized the SDWA for being unrealistic in the standards that are set. Financial and regulatory burdens placed on public water supplies are regarded as particularly unfair.[51]

A new program will have limited information at its disposal, and few resources to collect, analyze and interpret data. As a program matures, its database grows, as does its ability to present analysis. The level of information evolves from limited to comprehensive. Its ability to organize the process of rulemaking depends on its ability to marshall facts and figures to represent the outcomes of various decisions. It is desirable for an agency to maximize the amount of information to which it has access. This requires budgetary and institutional support.

Those who seek a more rational basis for environmental policy to set environmental quality standards have long been frustrated by the pluralistic process.[52] Rationality is thwarted by incremental, disjointed decisionmaking and meeting an agency clientele's demands. The lack of coordination between rulemaking and enforcement reduces the net benefits of environmental protection.[53] One paradox of pluralistic bargaining is that more rigorous analysis does not rationalize the standard setting process. Chronic data shortage can be used as a dilatory tactic. Requirement of complete answers to complex questions results in decisionless drift.[54]

No interest group would suggest that rationality is wrong. Without some technical criteria, the regulator has broad discretion in making rules. Business fears an agency so endowed will impose an unnecessary or unreasonable burden. Environmental groups fear industry capture of the regulatory mechanism. A consensus for technical decision making can evolve, depending on the way the technological criteria are framed. Both business and environmental interests fear the regulatory agency's a lack of accountability to the legislature.

SPECULATIVE AUGMENTATION

When political pressure to improve the environment demands immediate action, agencies responsible for setting standards cannot pursue standard-setting in a rational, analytical manner. Decision-makers seize a 'grand opportunity' to enact reforms when public opinion is clearly articulated, and rapid change in the regulatory regime is mandated. Charles O. Jones called this situation "Public satisfying speculative augmentation."[55]

In marked departure from the other models of standards setting, data are collected and analyzed after, rather than before standards are set. Scientists attribute risk perception to muckraking journalism and lurid sensationalism. Anecdotal evidence may be more informative and memorable to citizens than reams of statistics.[56] Exposés that successfully mobilize popular opinion around an issue makes it difficult for agencies to simply 'muddle through.' Demands for expedient action prevent data collection and analysis before the problem can be solved.

Speculative augmentation is marked by broad discretion given to a regulatory agency to deal with a perceived crisis. Reforms usually include strengthening the coercive power of the agency and the promulgation of detailed rules. Politicians must capitalize on public opinion, rather than wait for consensus. Opposition from the regulated industry, the scientific community and other private interest groups is unpopular. Regulatory agencies are then given the task of implementing this popular mandate before public attention is diverted and the regulated parties mount effective opposition. If the legitimacy of the technocracy is sufficiently undermined, rules are set in response to the electorate.

Environmental legislation passed under conditions of speculative augmentation has been notably difficult to implement.[57] The lack of data make any rules promulgated susceptible to charges of being arbitrary and capricious. Without a consensus of values and without easily quantified performance standards, achievements and progress under regulation make policy analysis inconclusive. Delay of standard setting and requirement of comprehensive risk assessment favors the *status quo*, and its supporters. Implicit in speculative augmentation is technology forcing. Cost of compliance data are irrelevant if the standards will bring about technological change to meet them.

Public attention must focus on an issue long enough to produce agency action. Crisis conditions are often part of an issue-attention cycle that produces political demands for immediate solution. The greatest amount of activity is associated with "alarmed discovery and euphoric enthusiasm."[58]

Once standards are set, the public will assume the problem has been solved. Problems with the new standards won't be identified until after they are implemented. Standards set by speculative augmentation may be inefficient and costly. However, to the extent they satisfy public demand they are optimal in the democratic sense. The Federal system and decentralized administration can channel pressure for the public. Drinking water became an issue after many water supplies became contaminated with hazardous wastes, agricultural chemicals, and petroleum products.[59]

The EPA relies heavily on state monitoring and enforcement of the SDWA. States are responsible for primary enforcement of federally promulgated regulations.[60] A water supply system that does not meet these standards must take whatever steps are necessary to bring the system into compliance at the earliest feasible time. States can impose their own MCLs as long as they are within the Federal limits. States have been notably inconsistent in implementation of the SDWA. To deal with the crisis, state governments have acted to set standards independently of EPA, rather than wait. Many of these state MCLs are based on Federal RMCLs. Many have set MCLs more stringent than the Federal levels, but monitoring and enforcement have fallen short of requirements in several states.[61] It takes far fewer resources to set a standard as a symbol than to enforce one as a law.

Industry groups prefer that MCLs be established at the Federal level, rather than face 50 sets of standards. However, the Chemical Manufacturers Association expressed fear that the standards set would not be "based upon sound scientific data."[62] Similarly, local governments did not want to be put in the position of trading jobs for safe water in competition with other regions. The length of delays in setting drinking water standards has led Congress to intervene.[63] The result was the Safe Drinking Water Act Amendments of 1986.[64]

The 1986 Amendments established a list of 85 contaminants and gave EPA three years to set the MCLs for them.[65] The new law also contains action forcing language that requires EPA to set MCLs for contaminants which "may have any adverse effect on the health of persons and which is known or anticipated to occur in public water systems."[66] Methods to establish MCLs are left to the discretion of EPA. The advisory board no longer sets RMCLs, but MCL goals, and EPA has greater responsibility in seeing these goals become MCLs.[67] President Reagan, upon signing the bill, questioned the statutory mandate requiring EPA to regulate specifically listed contaminants, saying it "seriously curtails the EPA administrators flexibility to determine which contaminants actually need to be regulated to protect the public health, and when . . . "[68] EPA is not given the explicit exemption from Executive Order 12291 in setting MCLs sought by environmental groups.[69] Instead, Congress required that cost be considered when setting MCLs.[70]

Speculative augmentation describes exceptional circumstances, yet accounts for regulatory activity not explained by rational or pluralistic behavior. Agency inertia fits the other models well; agency action seldom does. Standards set by speculative augmentation break the usual pattern of policy analysis by being initiated by democratic institutions rather than neutral experts or special interest groups. Speculative augmentation is not a mere catch-all for regulatory decision-making. To be legitimate, rules made by speculative augmentation require widespread popular support to overcome the inertia associated with other models of standard setting.

CONCLUSION

The SDWA failed to proactively address pesticide contamination. Instead, policy to protect drinking water was a reaction to an alarmed public faced with water sources that were already contaminated. Drinking water standards are designed to protect public health, but lack the ability to control the land uses associated with water pollution. One place were the public reaction was particularly strong was Long Island in New York State.

NOTES
CHAPTER IV

1. James W. Vaupel, "Truth and Consequences: Some Roles for Scientists and Analysts in Environmental Decisionmaking," in Wesley A. Magat (ed.) *Reform of Environmental Regulation*, (Cambridge, MA: Ballinger Publishing, 1982): 71-92.

2. P.L. 93-523 (1974), 42 U.S.C. §300 ff.

3. P.L. 93-523 §1412(a)(2) (1974); 42 U.S.C. §300g-1(a)(2) (1985). For legislative history on the intent of Congress and administrative implementation, see 1974 *U. S. Code Congressional and Administrative News* 6454; Congressional Research Service, *A Legislative History of the Safe Drinking Water Act*, (Washington, DC: U.S. Government Printing Office, 1982).

4. P.L. 93-523 §1412 (1974(), 42 U.S.C. §300g-1 (1985).

5. C. C. Johnson gives an abstract of the National Drinking Water Advisory Council transcript of 1977 meetings in "Drinking Water Policy Problems Background of the Current Situation," Clifford S. Russell (ed.), *Safe Drinking Water: Current and Future Problems.* (Washington, DC: Resources for the Future, 1978): 21-46.

6. 50 *Fed. Reg.* 46893ff. (November 13, 1985).

7. June Fessenden-Raden and Rita R. Calvo. "Aldicarb in Groundwater: Determining Acceptable Risk." *General Ecology's Water Research Update* 2(2) (1986).

8. Ronald Brickman, Sheila Jasanoff and Thomas Ilgen, *Controlling Chemicals: The Politics of Regulation in Europe and the United States.* (Ithaca, NY: Cornell University Press, 1985).

9. Rob Coppock, "Control of Chemical Hazards: Risk Analysis, Conceptual Outlooks and Political Traditions in Selected Countries." *Journal of Public and International Affairs* 5: 67-83 (1984).

10. Brickman, Jasanoff and Ilgen, *Controlling Chemicals:* pp. 58-59.

11. *Industrial Union Dept. of the AFL-CIO v. American Petroleum Institute*, 448 U.S. 607, 65 L.Ed. 2d 1010, 100 S.Ct. 2844 (1979); *American Textile Manufacturers Institute, Inc. v.*

Donovan, 452 U.S. 490, 69 L.Ed. 2d 185, 101 S. Ct. 2478 (1981).

12.   W. W. Lowrance, *Of Acceptable Risk*, (Los Altos, CA: Kaufmann, 1976).

13.   Talbot Page, *The Economics of Involuntary Transfers*. (Berlin: Springer-Verlag, 1973).

14.   Hal Varian, *Microeconomic Analysis*. (New York: W.W. Norton, 1978): 145-155.

15.   Buchanan, J. M and G. Tullock (1962) *Calculus of Consent*. Ann Arbor: University of Michigan Press.

16.   Yaron Ezrahi, "Utopian and Pragmatic Rationalism: The Political Context of Scientific Advice." *Minerva* 18: 111-131.

17.   P.L. 92-523 §1401(1) (1974); 42 U.S.C. §300f(1) (1986).

18.   *EDF v. Costle*, 578 F.2d 337, 344.

19.   P.L. 93-523 §1446 (1974); 42 U.S.C. §300j(5) (1986).

20.   *Legislative History of the Safe Drinking Water Act*: 550-552, 894.

21.   National Research Council, *Decision-Making in the Environmental Protection Agency*. (Washington: National Academy of Sciences, 1977).

22.   See Brickman, Jasanoff and Ilgen, *Controlling Chemicals*: 304.

23.   See, for example, the questioning of Congressmen Dannemeyer [R-CA] and Gramm [then D-TX] of Dr. Joe Cotruvo of EPA. U.S. Congress. House. Committee on Interstate and Foreign Commerce. Subcommittee on Health and the Environment. *Quality of Drinking Water--1980* (Washington, DC: U.S. Government Printing Office, 1980): 364.

24.   George Denning, economist at EPA, personal communication, 3 June 1986.

25.   Herbert Simon, "A Behavioral Model of Rational Choice," *Quarterly Journal of Economics* 69: 99-118 (1955).

26. Stahrl W. Edmunds, "Environmental Policy: Bounded Rationality Applied to Unbounded Ecological Problems." *Policy Studies Journal* 3(9): 359-369 (1980).

27. Herbert Simon, *Administrative Behavior*, (New York: Free Press, 1957): 182-186.

28. 44 *Fed. Reg.* 68641 (1979).

29. David G. Hoel and Kenny S. Crump, "Waterborne Carcinogens: A Scientist's View," in Robert W. Crandall and Lester B. Lave's *The Scientific Basis of Health and Safety Regulation*. (Washington, DC: The Brookings Institution, 1981): 185.

30. Hoel and Crump, "A Scientist's View:" 195.

31. Talbot Page, Robert Harris and Judith Bruser, "Waterborne Carcinogens: An Economist's View." in Crandall and Lave, *The Scientific Basis of Health and Safety Regulations*: 213-216.

32. Page, Harris and Bruser, "An Economist's View:" 220-228.

33. Victor J. Kimm, Arnold M. Kuzmack and David W. Schnare, "Waterborne Carcinogens: A Regulator's View," in Crandall and Lave's *The Scientific Basis of Health and Safety Regulation*: 229, 241-242.

34. Alfred A. Marcus, "EPA's Successes and Failures." in Kamieniecki, O'Brien and Clarke's *Controversies in Environmental Policy*: 153-173.

35. Daniel J. Fiorino and Patricia A. Wilbur, "Toward an Analytical Culture: Reflections on the Use of Policy Analysis in Environmental Decisionmaking." Paper presented at the American Society for Public Administration National Conference, Anaheim, CA (1986).

36. This evolution is marked by James Q. Wilson (1980), *The Politics of Regulation*. New York: Basic Books: 357-394 and is also presented in Douglas D. Anderson, "Who Owns the Regulators?", *Wharton Magazine* 4(4): 14-21 (1980).

37. John Kenneth Galbraith, *American Capitalism: The Concept of Countervailing Power*. (Boston: Houghton-Mifflin, 1952).

38. Theodore Lowi, *The End of Liberalism*. (New York: W. W. Norton, 1969).

39. Nicholas Watts. "From Consensus to Dissensus: The Role of Distributional Conflicts in Environmental-Resource Policy," in A. Schnaiberg, N. Watts, and K. Zimmerman, *Distributional Conflicts in Environmental Resource Policy*. (New York: St. Martins Press, 1986): 1-14.

40. The distributional conflicts of environmental quality are discussed in Lester Thurow, *The Zero-Sum Society*. (New York: Basic Books, 1980): Chapter 5.

41. Joseph P. Kalt and Mark A. Zupan, "Capture and Ideology in the Economic Theory of Politics." *American Economic Review* 74: 279-300 (1985).

42. Mancur Olsen, *The Logic of Collective Action*. (Cambridge, MA: Harvard University Press, 1965).

43. David F. Noble and Nancy E. Pfund, "Business Goes Back to College." *The Nation* 231 (September 20, 1980): 247-248. Noble and Pfund cite Bruce Owen and Ronald Braeutigam, *The Regulation Game*.

44. Walter Adams and James W, Brock, "Corporate Power and Economic Sabotage." *Journal of Economic Issues* 20: 919-940 (1986).

45. U.S. General Accounting Office, *Improving the Scientific and Technical Information Available to the Environmental Protection Agency in its Decisionmaking Process*. (Washington: U.S. Government Printing Office, 1979): 5.

46. See the testimony offered before the House by Seth Abbott of the American Petroleum Institute in *Quality of Drinking Water--1980*: 640-667.

47. Dickson and Noble explore in detail business associations' campaign against regulation and democratic institutions during the late 1970s in "By Force of Reason," 267-292.

48. Lowi, *The End of Liberalism*: 113-126.

49. Paul Sabatier, "Social Movements and Regulatory Agencies: Toward a More Adequate--and Less Pessimistic--Theory of 'Clientele Capture.'" *Policy Sciences* 6: 301-342, esp. 325-327 (1973).

50. Testimony and statement of Sam B. Dixon, Association of Metropolitan Water Agencies, before the U.S. Congress. Senate. *Safe Drinking Water Act Amendments of 1985*. (Washington, DC: U.S. Government Printing Office, 1985): 70.

51. Robert M. Clark and Richard G. Stevie, "Meeting Drinking Water Standards: The Price of Regulation" in Clifford S. Russell (ed.) *Safe Drinking Water: Current and Future Problems*. (Washington, DC: Resources for the Future, 1978).

52. For example, A. Myrick Freeman and Robert Haveman:

"Setting environmental quality standards is a meaningless exercise unless effective mechanisms are developed for achieving the standards . . . states have place primary reliance on some form of licensing of discharges . . . this leads to multi-stage bargaining with minimal accessibility and accountability, and the consequence is delay or non-enforcement of the standard."

"Clean Rhetoric and Dirty Water." *Public Interest* 28: 65 (1972).

53. Roland N. McKean, "Enforcement Costs in Environmental and Safety Regulations." *Policy Analysis* 6: 269-289 (1980).

54. Brian Wynne, *Rationality and Ritual*. (Chalfant St. Giles, UK: British Society for the History of Science, 1982): 32.

55. Charles O. Jones, "Speculative Augmentation in Federal Air Pollution Policy-Making," *Journal of Politics* 36: 449ff (1974).

56. Paul Slovic, Baruch Fischoff and Sarah Lichtenstein, "Facts versus Fears: Understanding Perceived Risk." in Daniel Kahneman, Paul Slovic and Amos Tversky, *Judgment Under Uncertainty*. (New York: Cambridge University Press, 1982): 464.

57. Henry C. Kenski and Helen M. Ingram, "The Reagan Administration and Environmental Regulation: The Constraint of the Political Market" in Kamieniecki, O'Brien and Clarke *Controversies in Environmental Policy*: 281-282.

58. Anthony Downs, "Up and Down with Ecology--The 'Issue-Attention Cycle." *Public Interest* 28: 38-50 (1972).

59. Many of the incidents are recorded in Fred Powledge, *Water*. (New York: Farrar, Straus, Giroux, 1982): 41-95, et seq.

60. P.L. 93-523 §1412(b)2(A) (1974); 42 U.S.C. 300g-2(a) (1985).

61. U.S. General Accounting Office. *States Compliance Lacking in Meeting Safe Drinking Water Regulations*. Washington, DC: U.S. Government Printing Office (1982).

62. Statement presented by Mitchell Eichelberger of the Chemical Manufacturer's Association before the U.S. Congress. House. Committee on Energy and Commerce. Subcommittee on Health and the Environment, *Environmental Issues*. (Washington, DC: U.S. Government Printing Office, 1985): 144-145.

63. Congressional Record, May 21, 1986: 6284-6301.

64. P.L. 99-339; 100 Stat. 642. Codified as 42 U.S.C. 300f *et seq.* Legislative history at 1986 *U.S. Code Congressional and Administrative News* 1566.

65. P.L. 99-339 §101(b)(1) (1986); 42 U.S.C. §300g-1(b)(1). The list is spelled out in 1986 U.S. Code Congressional and Administrative News *1593-1594*.

66. P.L. 99-339 §101(b)(3)(A) (1986); 42 U.S.C. 300g-1(b)(3)(A) (1987).

67. P.L. 99-339 §101(a)(2), §101(b)(2)(A) (1986); 42 U.S.C. 300g-1(a)(2), 42 U.S.C. 300g-1(b)(A) (1987).

68. 1986 *U.S. Code Congressional and Administrative News* 1614-1615.

69. See testimony offered by Jacqueline Warren of the Natural Resources Defense Council, in U.S. Congress. Senate. Committee on the Environment and Public Works. Subcommittee on Toxic Substances and Environmental Oversight. *Safe Drinking Water Act Amendments of 1985*. (Washington, DC: U.S. Government Printing Office, 1985): 76.

70. P.L. 99-339 §101(b)(5) (1986).

Long Island's Agriculture

Long Island[1] has long been an important agricultural region of the East Coast. Early settlers realized its potential for high agricultural productivity with its deep, easy to work soils and temperate climate. Potatoes and vegetable crops have done particularly well. However, in recent years, agriculture on Long Island has been retreating so that it now exists only on the easternmost part of the island. The principle reason for this has been the encroachment of suburbs, but other factors have included central city growth on the island and second home development in resort areas. To complicate matters further, Long Island agriculture is currently in a transition provoked by environmental problems. Chemicals used to control potato pests over some 26,000 acres have been found in Long Island's groundwater. This is the only source of drinking water for the over 2.6 million residents of the Island and represents 4.5% of the total recharge area.

Agricultural production on Long Island is concentrated on the eastern end of the Island. Most agricultural land is on the North Fork, consisting of Riverhead and Southold towns, with a substantial amount left on the South Fork, which include the towns of Southampton and East Hampton. These towns are shown in figure 5.1.

The character of Long Island's agriculture is far from typical of farming in the rest of the state. Farmers on Long Island face conditions which are different from farmers anywhere else in the nation as well. Hardy's exhaustive mid-century study recognized this.[2]

New York
Huntington
Smithtown
Brookhaven
Babylon
Islip
Riverhead
Southampton
Southold
Shelter Island
East Hampton

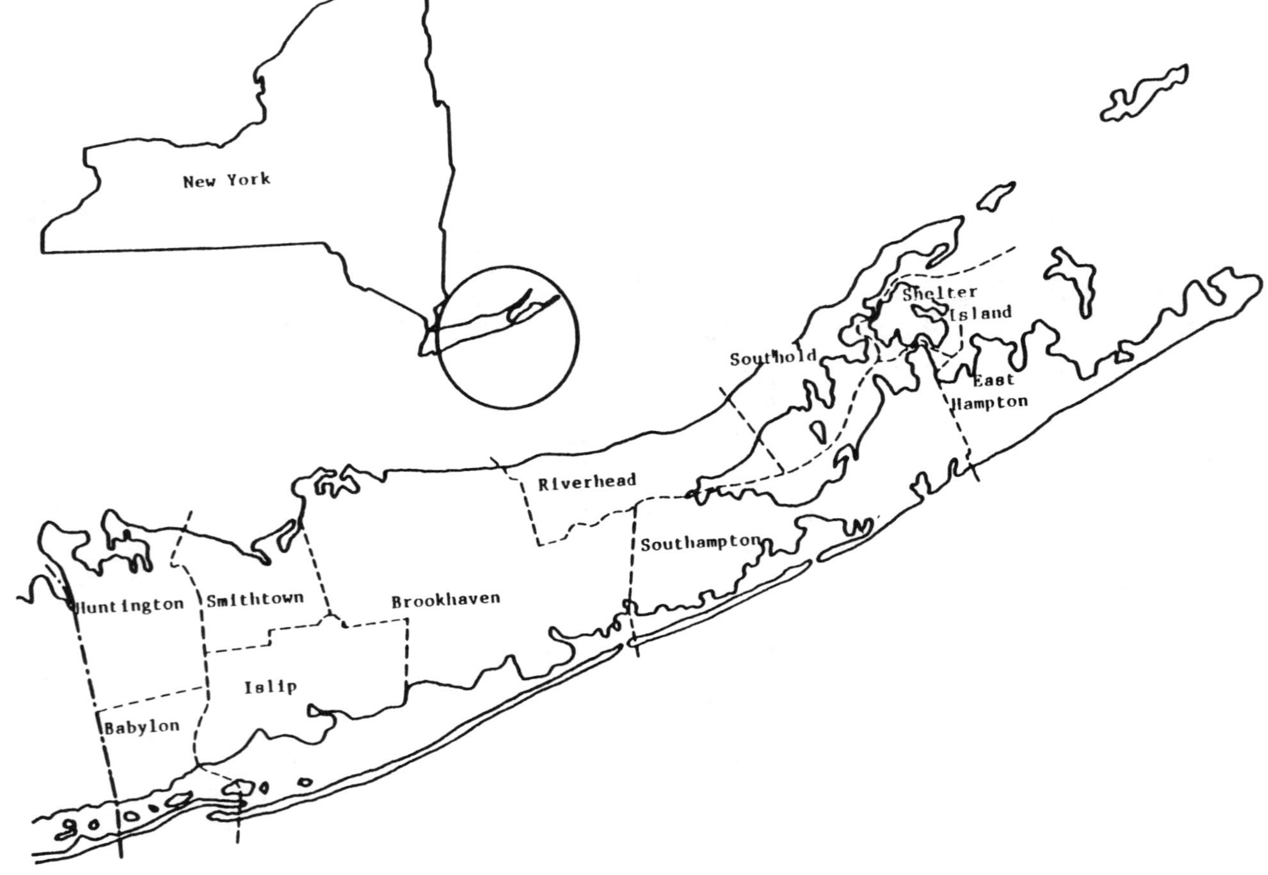

Figure 5.1

The agriculture of Long Island deserves special treatment in any study of New York's agriculture. There is no comparable area of high intensity agricultural production anywhere else in the state . . . (Other areas) do not produce the same kind of products under the same conditions of high priced land, high inputs per acre or complete marketing programs operated by farmers.

Long Island's cropland has long been dominated by potato production. Potatoes are still the most important crop in Suffolk County. Potato production is concentrated in five towns: Riverhead, Southold, Southampton, East Hampton and Brookhaven.

The topography, soil and climate of this area is highly uniform. Soil and climate vary slightly between the North and South Forks, and are well suited for potatoes. While production costs were and are higher on Long Island than in other parts of the Northeast, yields are much greater. The close proximity to the New York City market give Suffolk County a large market and low transportation costs.[3]

GROWING CONDITIONS

Two soil associations dominate Long Island agricultural land. These are the Haven-Riverhead sandy loam and Bridgehampton-Haven silt loam. Properly fertilized, Long Island soils are among the most productive in the state. The soils are deep, free of large stones, level, and well drained. They are light and easy to work. On the other hand, Long Island soil lacks organic matter and is low in nitrogen. Another disadvantage is that these soils have a low capacity for holding water and require a great deal of irrigation. Of potato land, about half is irrigated on the South Fork, and 75% is irrigated on the North Fork. Virtually all vegetable land is irrigated. The depth of soil varies somewhat, and this affects water holding capacity, as well as the susceptibility to leaching.

Long Island's climate is outstanding for the Northeast. The temperatures are moderate, not too hot in the summer, with as long a frost free period as anywhere in the state. While the Setauket (North Shore) and Bridgehampton (South Shore) weather stations record lower growing degree day season averages than many upstate stations, these degree days are spread over a longer period. The cooler temperatures are favorable for the production of potatoes and many vegetables. The earlier planting date gives Long Island potato farmers a competitive advantage over upstate, Maine and Canadian farmers who have to wait longer before planting in the spring.

Table 5.1

Long Island
Average Growing Degree Days Over 10° C.

Month	Bridgehampton	Setauket
March	1	4
April	24	49
May	101	151
June	230	274
July	408	452
August	320	343
September	241	267
October	120	155
November	22	30

Source: Converted from fahrenheit to centigrade from the table for 50° fahrenheit in B. E. Dethier and M. T. Vittum, *Growing Degree Days in New York State*, (Ithaca, Cornell University Agricultural Experiment Station, 1967):40, 48.

MARKETS

An important historical advantage Long Island agriculture has had over other areas has been close proximity to the largest market in the country: New York City. In the 18th and early 19th centuries, Long Island products were able to be shipped to New York City by water and rail before canals built gave upstate farmers access to the lucrative New York market. In addition to the New York City market, there is a substantial market on Long Island itself.

Suffolk County leads New York State in the production of potatoes, cauliflower, duckling, cut flowers, sod, turf, nursery stock, and clams. The county is also has the highest gross farm receipts in the state.[4] Potatoes and cauliflower have developed highly specialized marketing networks on Long Island. In the case of potatoes, farmers usually harvest, grade and pack the potatoes. These are then wholesaled to broker who transports them to a terminal market. The principal terminal markets for Long Island potatoes are New York-Newark, Boston, Philadelphia, Baltimore-Washington, Atlanta, Pittsburgh, Providence, and Albany. These markets account for 87% of the potatoes grown on Long Island in 1983, with New York-Newark alone accounting for over one-third.[5]

Suffolk County cabbage and cauliflower are sold in most of the same markets as Long Island potatoes. Other Suffolk County vegetables had difficulty expanding into terminal markets during the 1960's and '70's. The trend since the late '70's, however, has seen a general increase in demand for Suffolk County vegetables.[6] Direct marketing has grown the fastest of all outlets for Suffolk County products.

AGRICULTURAL LAND

Long Island agricultural land is intensively utilized. The high opportunity cost of idle land discourages non-productive uses. High land values reflect the development pressure Long Island faces. Nassau County was once a major farm area, with the Hempstead Plains an important grain and livestock region. There are now no commercial farms in the two townships that make up the Plains; these towns now house a suburban population of close to one million.[7] Nassau County agricultural land decreased by 85% in the decade 1950-1960, and went from having the second most productive agriculture of any county in the state to nearly last.[8] Western Suffolk County met a similar fate. A few specialty farms growing such crops as turf and ornamentals remain in Nassau and Western Suffolk Counties. The loss of land in agricultural production is shown in figure 5.2.

The main reason for the loss of Long Island farmland is suburban sprawl. The development of the automobile and limited access highways enabled people to commute to New York City by car on a daily basis. Land went out of agriculture and into subdivisions. Expressways were built through some of the best agricultural land on Long Island because they were the path of least resistance.[9]

Suffolk County Land in Farms
1954-1984
Acres
(Thousands)
100
90
80
70
60
50
40
30
1954
1959
1964
1969
1974
1979
1984
Year

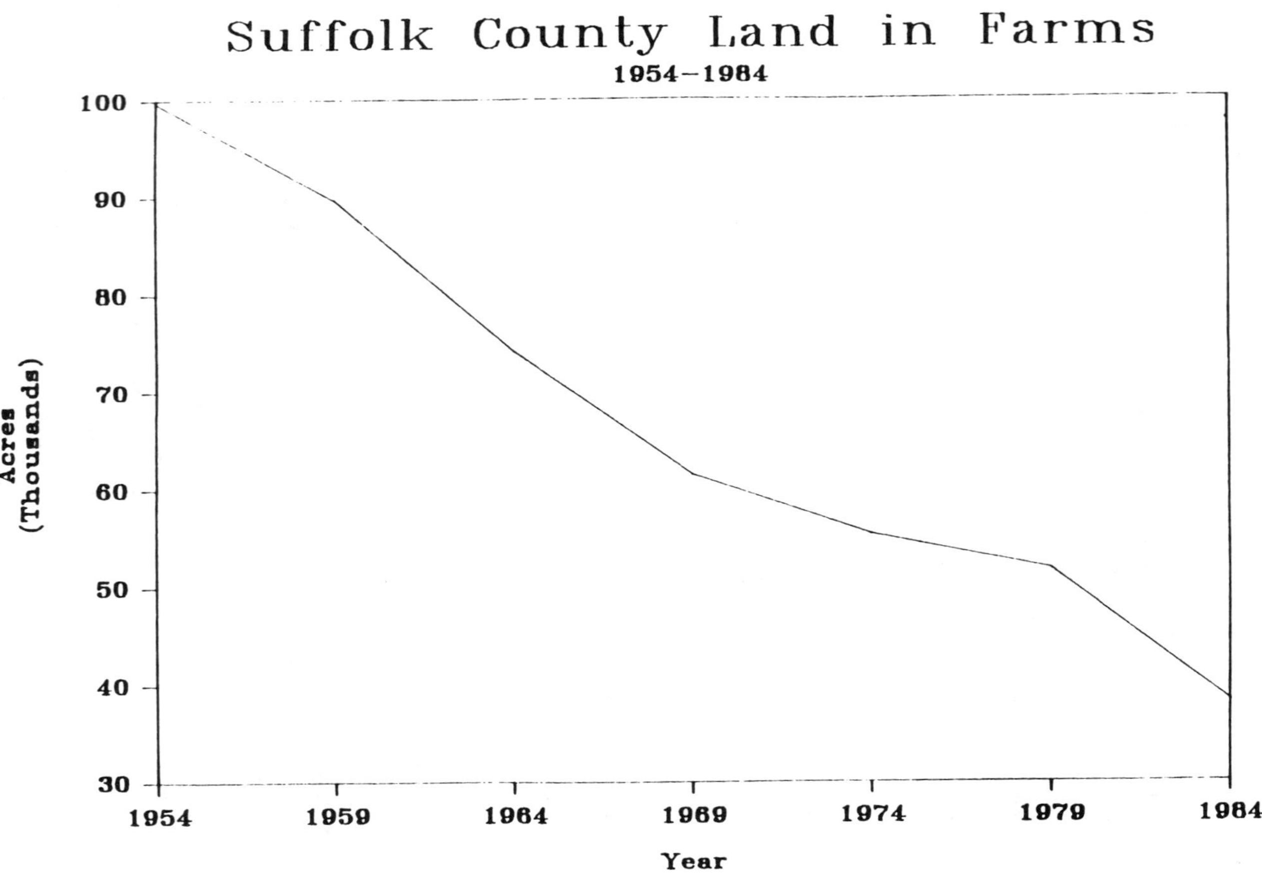

Figure 5.2

The fragmentation of land tenure has, in some cases, made potato production uneconomical. With increased mechanization of potato production, the scale of these operations have increased. Two factors mentioned above, development of land for residential and vacation uses, and intensive vegetable production, have led to a decrease in the average farm size on Long Island. (See table 5.2).

Table 5.2
Average Farm Size
Suffolk County, New York
1954-1982

Year	Farm Size (acres)
1954	68.1
1959	71.4
1964	65.3
1969	82.7
1974	75.0
1979	67.0
1982	61.5

Source: U.S. Dept. of Commerce, Bureau of the Census, *Census of Agriculture*. Various years.

As figure 5.3 shows, potato acreage has been declining at a rate more rapid than all agricultural land in recent years. There has been a tendency for potato farmers to sell their land to developers. These developers often keep some land in agriculture by renting to intensive small scale vegetable enterprises. The remaining land is sold off as house lots. Because land sold is seldom immediately developed, much agricultural land is rented to farmers. A 1983 survey showed that almost 60% of all land in agricultural production was rented.[10]

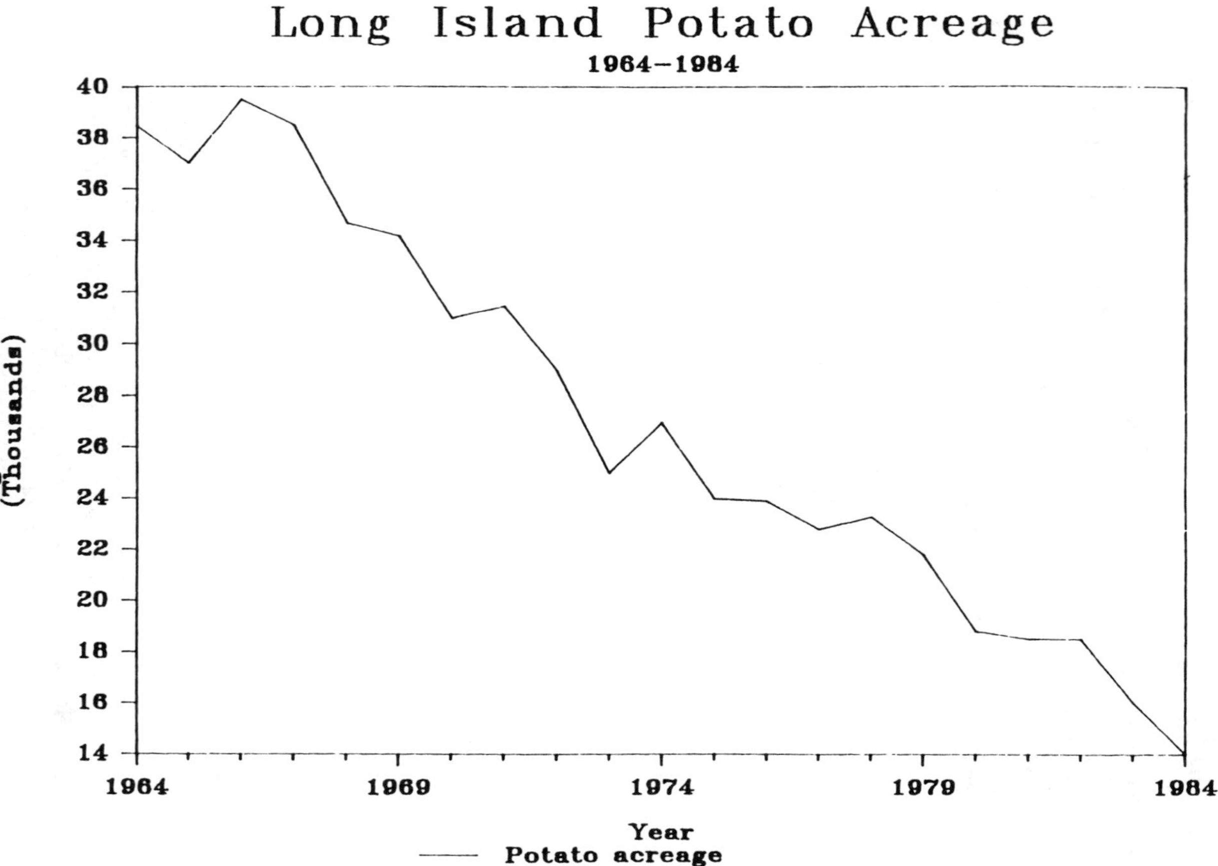

Long Island Potato Acreage
1964-1984
Acreage in Potatoes
(Thousands)
40
38
36
34
32
30
28
26
24
22
20
18
16
14
1964
1969
1974
1979
1984
Year
Potato acreage
Figure 5.3

Small grains are grown as cover crops, either in rotation with potatoes, or as a groundcover in the interim period between commercial production and conversion to non-agricultural use. Rye is also harvested for a strong local market. Cole crops, particularly cauliflower, have a long tradition of cultivation on Long Island.

The climate and soils of Long Island make it more suitable for diversification than many other parts of the Northeast. A wide variety of crops are already harvested, but with significantly less acreage than potatoes.

Research on the economic feasibility of alternative crops has been performed at Cornell University.[11] Sweet corn is frequently marketed on roadside stands. Other vegetable crops, are grown in quantities under 1,000 acres. The area has also had a stable acreage of strawberries over the years. Other fruit production, particularly grapes and peaches, has been increasing. Nursery operations, wine vineyards, and horse farms have increased. Economic forces have already begun a trend toward these alternative crops. The limiting factors for fruits and vegetables are the availability of labor and marketing institutions. If these factors can be overcome, then Suffolk County farmers could profitably diversify. The cropland mix over the 1982-1984 period is presented in table 5.3.

Table 5.3

Crop Acreage on Long Island

Crop	1982	Year 1983	1984
Potatoes	18,500	16,000	14,000
Rye and wheat	4,000	4,000	4,000
Cauliflower	2,000	2,000	2,000
Cabbage	1,700	1,600	1,500
Sweet Corn	1,200	1,200	1,200
Snap Beans	500	500	500
Cucumbers	300	300	300
Spinach	300	300	300
Onions	300	300	300
Peppers	300	300	300
Tomatoes	200	200	200
Lettuce	200	200	200
Squash	200	200	200
Pumpkins	200	200	200
Grapes	500	700	900
Peaches	400	450	500
Apples	330	330	330
Strawberries	300	300	300
Total	31,430	29,080	27,230

Source:William Sanok, Suffolk County Cooperative Extension, personal communication, and Ron Leuthardt, Suffolk County Farm Bureau, personal communication.

The market for agricultural land in Suffolk County is unlike any other land market in New York State. Nowhere else is development pressure as great; land speculation remained strong on Long Island, particularly on the South Fork, well into the eighties after many of the other land market bubbles burst during the decade. The main reasons for this pressure on the land market stemmed from population growth, tourism and second home development. Suffolk County was the fastest growing county in New York State during the decade from 1970-1980. After most other areas on the urban fringe saw population growth level off, Suffolk County's rate was an explosive 41%.[12]

The tourist industry attracted a large volume of business throughout the 1970s, and continues to be the mainstay of the eastern part of Long Island's economy. The 1970s also saw eastern Long Island, particularly the South Fork, become a chic vacation spot for many rich, young professionals. Condominium unit prices soared, land development accelerated, and some questioned the South Fork's ability to maintain its identity as a resort.[13] Speculative activity by farmers, the largest land owners on the island, was expected to wipe out agriculture in a few years, as it had done in Nassau and western Suffolk Counties in the decade between 1950 and 1960.

Suffolk County has experimented with, or at least considered, almost every type of program for preserving agriculture. The success of these programs is relative to what would have happened in their absence, something nobody can know for certain.

The farmers and citizens of Suffolk County have put a great deal of effort into preserving agriculture and agricultural land in Suffolk County. There has been some debate over the threat to agriculture, and the best way to preserve agricultural land. After the first wave of suburban expansion, one writer argued that Long Island's agriculture is strong enough to withstand any take-over by speculators for some time to come.[14] More recent observers point to the rapid depletion of agricultural land in Nassau and Western Suffolk Counties and suggest that the same thing could happen to Eastern Suffolk.[15]

Despite efforts by Long Island's local governments to stop further loss of agricultural land, land continues to be converted to non-agricultural uses. These efforts have concentrated on zoning, the purchase of development rights (PDRs), the formation of agricultural districts, restrictive and planning. Local governments have also discussed and studied transfer of development rights (TDRs).

One frequently cited motivation for preserving agricultural land on Long Island has been environmental preservation. One environmental amenity that agriculture was meant to protect was groundwater quality.[16] The rationale behind this was undermined by the discovery of pesticides in Long Island groundwater.

PESTICIDE CONTAMINATION

The reliance on a potato monocropping system has caused Long Island agriculture to face an ecological problem. This monocrop system has been plagued by pest problems, specifically from the golden nematode and Colorado Potato Beetle. The soils and climate which make Long Island an ideal place for growing potatoes also cause pesticides to be transported through soils more rapidly than many other places: Porous soils and abundant rainfall. The sensitivity of Long Island's groundwater resources is well known. Nassau-Suffolk was the second aquifer in the country to be accorded sole source aquifer status, one of only 17 such designated aquifers in the country.[17] Chemicals used to control these potato pests have been found in Long Island's groundwater, the sole source of drinking water for the island's population of over 2.6 million. Two of the chemicals found, aldicarb (trade name Temik) and carbofuran (trade name Furdan), have had their registration for use on Long Island removed. A third, oxamyl (trade name Vydate) has been removed from Long Island by its manufacturer.

The towns of Brookhaven, Riverhead, Southold, Southampton and East Hampton have had all wells within 2,500 feet of potato fields tested for the pesticide aldicarb. Nearly 8,000 wells were tested, with concentrations ranging from not detectable to 484 parts per billion (ppb). Over 2,000 wells had detectable concentrations and, of those, over 1,000 wells had concentrations exceeding the state's recommended guideline of 7 ppb. This standard was set conservatively at 0.001 of the no effect dosage level.

Union Carbide, manufacturer of aldicarb, offered free installation of water filters to households with concentrations exceeding the state maximum contaminant level of 7 ppb. In return, homeowners had to agree to recharge and maintain those systems. Almost every household eligible accepted the offer. In spite of the conservative standard, this is evidence of risk averse behavior by consumers. Those who have wells with contamination levels below 7 ppb must pay for any treatment system installed.

There is evidence that aldicarb contamination is following a pattern of moving from shallow areas of high concentration into deep areas of low concentration.[18] This is caused by four major processes: Movement of aldicarb residues from the unsaturated zone into the aquifer; movement of aldicarb residues through the aquifer; dispersion of the aldicarb residues occurring in both the unsaturated zone and in the aquifer; and the degradation of aldicarb residues.[19]

Aldicarb is of particular concern because of its high acute toxicity, neurological damage from cholinesterase inhibition, and a steep dose-response curve.[20] One of the scientists who helped to register aldicarb stated that it could not be safely used in agriculture.[21]

While affected areas showed lower levels, unaffected areas began to show signs of contamination. Still, the level of contamination of the newly affected areas is lower than the contamination levels of the areas showing the first signs of contamination.[22] As the concentration of aldicarb decreases, other contaminants become of greater concern. The appearance of these pesticides in water well samples lead to action by the Suffolk County Department of Health Services and the New York State Department of Environmental Conservation.[23] Pesticides have been banned, contaminated public wells have been closed, and citizens have been advised where they are drinking water out of contaminated wells.

Farmers responded to the removal of these pesticides from the market in several different ways. Some switched to growing other crops. Some got out of farming completely. Most substituted other pesticides for the ones that were banned. Those farmers who are changing their production are switching to high value fruit and vegetable crops. The conversion to these crops will require substantially more labor than does potato production. This situation has forced farmers to consider alternatives to these specific pesticides. More importantly, it has forced farmers to examine pest control methods which do not use pesticides.

NOTES
CHAPTER V

1. The region referred to as Long Island in this study consists of Nassau and Suffolk County, and does not include the two boroughs of New York City which are geographically also on Long Island, Kings (Brooklyn) and Queens.

2. Ernest Hardy, *The Agricultural Regions of New York State: A Mid-Century Description.* (Ithaca, NY, Cornell University unpublished Ph.D. dissertation, 1969): 875.

3. For a historical comparison of potato farm production costs, receipts and incomes, see Floyd W. Underwood, *Factors Affecting Costs and Returns in Producing Potatoes in New York in 1929.* (Ithaca, Cornell unpublished Ph. D. Dissertation, 1932).

4. New York State Department of Agriculture and Markets, *Agricultural Statistics 1982* (Albany, New York State, 1984).

5. Gabriella de la Mora, *Long Island Potato Markets and Markets for Alternative Vegetable Crops.* (Ithaca, NY: Cornell University Unpublished M.S. Thesis, 1985): 68.

6. de la Mora, *Long Island Potato Markets*: 103-105, *et seq.*

7. The combined population of the towns of Hempstead and North Hempstead in the 1980 census was 957,141.

8. U.S. Department of Commerce, *Census of Agriculture, New York, 1959.* (Washington, US Government Printing Office, 1959): 133. Hardy, *New York Agriculture at Midcentury*: 874.

9. Robert Caro, *The Power Broker* (New York, Vintage Books, 1974): 279-280.

10. Gerald White, David Lee and Ken Gardner, "Survey of Farm Businesses in Suffolk County", *Suffolk County Agricultural News* September, 1983: 3-4.

11. Mildred E. Warner, "Enterprise Budgets for Potatoes, Wheat, Cauliflower, Peaches and Table Grapes on Long Island, New York: A Comparison of Costs, Returns and Labor Requirements." (Ithaca, NY: Cornell University Dept. of Agr. Econ., A. E. Res. 85-12, 1985); S. S. Lazarus and G. B. White, "The Economic Potential of Crop Rotations in Long Island Potato Production." (Ithaca, NY: Cornell University Dept. of Agr. Econ., A. E. Res. 83-20, 1983).

12. U.S. Department of Commerce, Bureau of the Census, *Census of the Population--New York.* (v. 34). (Washington, DC, US Government Printing Office, 1980).

13. Paul Goldberger, "The Strangling of a Resort", *New York Times Magazine* September 4, 1983: 14 ff; Henry Post, "Forget the Hamptons, Now It's Country Chic", *New York* August 10, 1981: 24 ff.

14. Hardy, *New York Agriculture at Mid-century*: 876.

15. Johan B. W. Scholvinck, *Preserving Agriculture on Long Island.* (Ithaca, Cornell University unpublished thesis, 1974): 23 ff.

16. Scholvinck, *Preserving Agriculture on Long Island*: 8-12.

17. As of July, 1984. U. S. Congress, Office of Technology Assessment, *Managing the Nation's Groundwater* (Washington, DC: U.S. Government Printing Office, 1984): Vol. II, p. 225.

18. Henry B. F. Hughes and Keith S. Porter, *Interim Results Tracking Aldicarb Residues in Long Island Ground Water* (Ithaca, Cornell University Center for Environmental Research, 1984).

19. Hughes and Porter, *Interim Results*: 26.

20. A summary of the health effects of aldicarb is contained in Dave Bunn, *The Case Against Aldicarb.* (Los Angeles, CA: Pesticide Watch, 1990).

21. Elliot Marshall, "The Rise and Decline of Temik," *Science* 229: 1369-1371 (1985).

22. Hughes and Porter, *Interim Results*: 28.

23. Joseph Baier and Sy Robbins, "Report on the Occurence and Movement of Agricultural Chemicals in Groundwater" (Hauppauge, NY: Suffolk County Department of Health Services, 1982).

A Model of Long Island's Agriculture

To examine alternative policies for protecting groundwater, a model of Long Island's agriculture is constructed. Models are abstractions from reality based on simplifying assumptions so that a logical conclusion can be drawn that answers, in a limited sense, the question of interest. Specifically, what are the economic and environmental effects of different policies to protect groundwater on Long Island? To answer this question, the model must integrate economic decisionmaking, pest pressure and groundwater quality.

While models need not be mathematical, such models are often used by economists because of the rigorous results that they produce. With the aid of computers, mathematical models are useful for the synthesis of diversified facts and formulae into a single logical foundation. A model should be simple enough to be easily interpreted, yet sophisticated enough to reflect adequately the nature of the problem.

ECONOMIC MODEL

The basic economic model is a recursive, stochastic programming model (RSP).[1] The basis of RSP is linear programming or LP.[2] LP is a normative tool. Rather than saying what farmers produce or what farmers will produce, an LP model tells farmers what they ought to produce to maximize profits. LP also has uses in positive economics if one can assume that economic actors behave according to the assumptions stated above.

The shortcomings of LP as a model for economic behavior should also be mentioned. LP is a single objective model. This implies that, even if an economic agent has more than one goal, only a single value is used to make a final decision. This is a feature of all optimization models, goal and multiple objective programming notwithstanding. LP is deterministic. Any problem which is subject to random disturbances cannot be fully reflected by an LP model.

The basic economic programming model is formulated at the microeconomic, or farm level. Micromodels are useful in examining production response for a single farm, but are not appropriate for examining policy changes which affect an entire region. The diverse impacts of policy changes cannot be wholly captured in a micromodel. Many of the implications of the policy may not hold for any given individual farm, but would be true for the region as a whole. To account for these shortcomings, a regional, rather than an individual farm model is used in the analysis of policy.

The regional model that follows is a scaled up version of components of these models. Again, the basis of this model is linear programming. However, there are some differences with this model which require some explanation. One is the problem of error introduced by aggregating models for individual farms. This error is known in the literature as aggregation bias. There has been substantial research on the different types of bias introduced by using micro models to analyze macro behavior. There have been a few reviews of the literature.[3] LP is usually used as a static model, but can be used to solve dynamic problems by recursive programming (RP).

RP solves a series of LP problems over several time periods where constraints and coefficients of the current period are a function of activities in the previous period. Adjustment to resource policies is usually adaptive in nature.[4] RP uses constrained maximization to describe the intended behavior of an economic agent or group of agents. Decisions reflect the accumulation of experience over time. Uncertainty leads to cautious behavior on the part of agents. Decisions in one period are viewed as deviations from decisions in the previous period. Behavioral rules are followed which make allowances for future decision-making, and those that modify objectives on the basis of past behavior, thereby limiting change.[5] Institutions that regulate agricultural impacts on groundwater face uncertainties that have impaired program development.[6] By introducing coefficients that change randomly, the model is then made stochastic and reflects the probabilistic nature of decisionmaking.

RSP is an appropriate analytical tool for researching the economic policies proposed to remedy agricultural pollution. The policies would, by their very nature, lead to structural change which renders statistical observations obsolete and irrelevant. There are a number of fixed factors in production which prevent the immediate adjustment to the new policies. A programming model which failed to account for certain rigidities in production response would have solutions that fluctuate wildly as a result of policy change.

Simulations are conducted over a five year time horizon. Resource allocation in the model is based on recursive linear programming. The objective function of the model is

$$(6.1) \quad \text{MAX } E[\Pi_t] =$$

$$\sum_{i \in I} \sum_{k \in K} p_i\, y_{ik} - \sum_{i \in I} \sum_{k \in K} c_{ik}\, x_{ik} -$$

$$\sum_{\psi \in \Psi} \sum_{k \in K} c_\psi\, Z_{\psi k}$$

$$\{t = 1, \ldots 5\}$$

Farmers are assumed to be risk neutral where $E[\Pi_t]$ is the expected profit in year t, p_i is the price of crop i, y_{ik} is the yield per acre of crop i on land k; c_{ik} is the variable cost per acre of producing crop i on land k (not counting pesticide variable costs); x_{ik} is the number of acres of crop i grown on land k; c_ψ is the cost of pesticide ψ; and $Z_{\psi k}$ is the amount of pesticide ψ used on land k.

Land categories k are divided according to fertility, climate, irrigation, and cropping history: continuous potato, potato/grain rotation, potato/ vegetable rotation, or continuous vegetable. Rotations are recursively linked. Activities included potatoes; field crops such as rye, wheat, oats, soybeans, field corn, dry beans and sunflowers; and vegetables, including cabbage, cauliflower, cucumbers, lettuce, onions, peppers, snap beans, spinach, sweet corn, tomatoes, summer and winter squash. The model has two sets of

$$(6.2) \quad \text{s.t.} \quad \sum_{i \in I} \sum_{k \in K} A_{ik} X_{ik} \leq B_{jt}$$

$$\{ \forall j \in J \; ; \; t=1, \dots 5 \}$$

Data from farm budgets were used for the objective function and technical coefficients.[7] The fixed constraints included family and hired labor by season.[8] Labor is a seasonally binding constraint that inhibits expanded cultivation of high-value, labor intensive crops, such as cauliflower and lettuce. It is perhaps the main factor that inhibits cultivation in the region.[9] Transfer rows are used to link production and market activities.

Land available for cultivation is determined by a set of flexibility constraints for the different crops. This models the adjustment process. Flexibility constraints for each crop are calculated by multiplying the upper and lower flexibility coefficients, $\bar{\beta}$, and $\underline{\beta}_i$ respectively, by the amount of crop i grown in period t-1. Flexibility constraints are given in (6.3 and 6.4).

$$(6.3) \quad \sum_{k \in K} X_{ikt} \leq (1+\bar{\beta}_i) X_{ik(t-1)}$$

$$(6.4) \quad \sum_{k \in K} X_{ikt} \geq (1+\underline{\beta}_i) X_{ik(t-1)}$$

Acreage changes from year to year are constrained by flexibility coefficients. The constraints were set by maximum historic changes. Flexibility coefficients used in the model are contained in table 6.1.

Table 6.1
Flexibility Coefficients

Crop	$\bar{\beta}$	$\underline{\beta}$
Beans	1.12	0.76
Cabbage	1.30	0.87
Cauliflower	1.27	0.86
Cucumbers	1.17	0.86
Field Corn	2.00	0.80
Grains	1.57	0.67
Lettuce	1.17	0.86
Onions	1.07	0.89
Peppers	1.09	0.79
Soybeans	3.00	0.00
Spinach	1.17	0.81
Squash	1.17	0.86
Sunflowers	3.00	0.00
Tomatoes	1.09	0.81

Source: Derived from New York State Dept. of Ag. and Markets, *New York Agricultural Statistics*, 1962-1982; William Sanok, Suffolk County Cooperative Extension, Personal Communication.

PLANT/PEST INTERACTION

To model the response of farmers to policies which change their decisions to apply pesticides, one must take into account the way that these changes affect pest population dynamics. Three things are relevant to economic decision making and environmental quality when modelling plant/pest interactions: quantity of sprays, timing of sprays and pest population effects on yield. With programmed spraying, quantity and timing of sprays are set *a priori*. Yields vary randomly with population density. As pest population density increases, pest damage to crops increases and this damage decreases the final yield of the crops.

Yields of potatoes are a stochastic function of CPB population density. Farmers adapt to increased CPB resistance to pesticides by changing pest control strategies. Adult pest densities have been shown by Logan to be a function of degree-days. The functional form used by Logan (6.5) was estimated for Long Island data collected in 1981 and 1982 using a Maximum Likelihood technique. Non-linearities prevented use of ordinary least squares (OLS) estimation.

$$(6.5) \qquad CPB = e^{\,b_0 + b_1(°D) + b_2(°D^2) + b_3(°D^3)}$$

where $°D$ is the number of degree days at $10°$ centigrade, and b_0, b_1, b_2, and b_3 are coefficients. The coefficients were adjusted for climate, soil type, irrigation practices, and rotations. Estimates of the coefficients are contained in Table 6.2.

These coefficients were used to simulate beetle populations for subsequent runs of the model. The variable $°D$ was randomly generated using Monte Carlo methods. The resulting predicted CPB populations were used to simulate integrated pest management (IPM) activities, such as scouting and timing of applications. Foliar pesticides were assumed to be applied to potatoes only when CPB populations exceeded the threshold (CPB^*) for a given field. Farmers were assumed to use the thresholds provided by Cooperative Extension for CPB^*.[10] Scouting took place in two week intervals over the season from May to September.

Table 6.2

Potato Beetle Growth Rate Coefficients

Field	b_0	b_1	b_2	b_3
1	-0.20E+00	0.12E-01	-0.40E-04	-0.60E-07
2	-0.20E+00	0.11E-01	-0.40E-04	-0.65E-07
3	-0.20E+00	0.12E-01	-0.40E-04	-0.60E-07
4	-0.20E+00	0.11E-01	-0.40E-04	-0.65E-07
5	-0.25E+00	0.12E-01	-0.40E-04	-0.60E-07
6	-0.25E+00	0.11E-01	-0.40E-04	-0.65E-07
7	-0.25E+00	0.12E-01	-0.42E-04	-0.60E-07
8	-0.25E+00	0.11E-01	-0.42E-04	-0.65E-07

Source: Adapted from Patrick A. Logan, "Estimating and Projecting Colorado Potato Beetle Density and Potato Yield Loss" in *Advances in Potato Pest Management*.

Field	Description
1	North Fork Irrigated Potato
2	North Fork Unirrigated Potato
3	North Fork Irrigated Field
4	North Fork Unirrigated Field
5	South Fork Irrigated Potato
6	South Fork Unirrigated Potato
7	South Fork Irrigated Field
8	South Fork Unirrigated Field

The model gave farmers a selection of methods to control the CPB. The decision to spray is represented by the following conditional expression

$$(6.6) \qquad Z_{\psi k \tau} = \begin{cases} \alpha_\psi & \text{if } CPB_{k\tau} \geq CPB^* \\ \\ 0 & \text{if } CPB_{k\tau} < CPB^* \end{cases}$$

where $Z_{\psi k\tau}$ is the loading rate for pesticide ψ on land k in scouting period τ and α_ψ is the recommended application rate for pesticide ψ.. This relation assumes both perfect information from scouting and perfect compliance with label instructions. When pesticides are applied, a certain percentage of the beetles will be killed. This percentage, known as the mortality rate, is a function of the size of the dosage, the toxicity of, and the resistance in the insects to the pesticide being used. The rate of survival can be thought of as

$$(6.7) \qquad S_{\psi\tau} = CPB_\tau(1 - M_\psi)$$

$$\{0 \leq M \leq 1\}$$

where $S_{\psi\tau}$ is the survival rate of CPBs treated with pesticide ψ in scouting period τ, and M_ψ is the mortality rate for CPBs to pesticide. The insects then recover from $S_{\psi\tau}$ and continue their growth. This model assumes that growth rates are unchanged by insecticide application within a season and remain affected only by time and temperature.

$$(6.8) \qquad M_{\psi t} = \begin{cases} \dfrac{M_{\psi(t-1)}}{R_\psi} & \text{if } Z_{\psi(t-1)} > 0 \\[2em] M_{\psi(t-1)} & \text{if } Z_{\psi(t-1)} = 0 \end{cases}$$

The mortality rate declines over the years as the insects become resistant to a given pesticide. This introduces a recursive aspect to the model, requiring a feedback loop that accounts for the declining efficacy of a pesticide. This states that for a given year, t, the mortality rate associated with a given pesticide will be a function of whether or not that pesticide was used the previous year. R_y is the resistance factor of pesticide ψ. This relationship is shown is (6.8). The coefficients used for resistance and mortality are contained in table 6.3.

Table 6.3
Mortality Rates and Resistance Factors
For the Various Pesticides

Pesticide	Mortality Rate	Resistance Factor
Aldicarb	0.99^2	0.75^1
Carbofuran	0.93^2	0.71^2
Kryocide	0.93^3	0.75
Oxamyl+Thiodan	0.87^3	0.68
Pydrin+PBO	0.90^3	0.77^1
Rotenone	0.97^3	0.80

Sources:

[1].Andrew J. Forgash, "Insecticide Resistance of the Colorado Potato Beetle, *Leptinotarsa decemlineata* (Say), in Lashomb and Casagrande (eds.) *Advances in Potato Pest Management*.

[2]Norman L. Gauthier, Richard N. Hofmaster, and Maurie Semel, "History of Colorado Potato Beetle Control" in Lashomb and Casagrande, *Advances in Potato Pest Management*.

[3]Maurie Semel, "1983 Insecticide Tests", unpublished paper of the L.I. Horticultural Research Laboratory, Riverhead, NY.

Predicted potato yields were calculated from cumulative CPB densities over the season, A. Densities are estimated by an Euler approximation of the integral,

$$(6.9) \qquad A_\tau = \int_0^{\overline{°D_\tau}} CPB(°D)d°D$$

A linear function estimated by Logan was used to capture the plant/pest interaction. The function,

$$(6.10) \qquad \hat{Y}_k = \bar{Y}_k - \sum_{\tau \in t,} \gamma A_\tau$$

where Y_k is the predicted yield of potatoes on field k, and $_k$ is the maximum potential yield of field k, γ is the slope of the plant/pest interaction estimated by Logan to be 0.000168; and A_t is the cumulative number of degree days in period τ. The expression in the brackets explains crop damage as a function of CPB population feeding over the growing season. Populations are larger and more active during warmer summers.

The pesticide loading rate for a field in a given year is derived by summing over scouting periods in the season (6.11).

$$(6.11) \qquad Z_t = \sum_{\tau \in t} Z_\tau$$

The computer model consisted mainly of a FORTRAN program which read output from MPSX, a linear programming package, formulated constraints and technical coefficients based on the simulation of pest population and the flexibility coefficients. These data were then written to a file which was read into MPSX, which solved a year's LP problem.[11]

The basis and solution for a base year were stored in separate files before initiating a run. The base year chosen for this study was 1983. A run consisted of simulating five growing seasons. Each growing season was broken up into two week scouting periods. Pesticide applications were stored for each of the two week scouting period in a season for the different crops. Other data stored during a run were the gross margins, acreages of different crops, dual variables for selected constraints, pesticide applications for potatoes and yields of potatoes.

Degree days were randomly generated using a subroutine from the International Mathematics Subroutine Library (IMSL), and the weather data cited in chapter 5. From these degree days, Colorado Potato Beetle populations were derived, and yield was derived from Colorado Potato Beetle populations. Each run was replicated 15 times to generate a distribution for each parameter in the model. The stochastic variables in the model were those variables affected by the randomly generated pest population, foliar application of pesticides to potatoes and potato yields. All other variables were deterministic and were assumed to be held constant throughout the period analyzed.

Cropland was assumed to decline at roughly the same rate as had been the recent historical trend over the recent past, approximately 4.5% per year. Potato and field land were assumed to leave agriculture at above the average rate, vegetable land was assumed to leave agriculture at below the average rate. The model allocated the sale of land by location, with the South Fork acreage declining more rapidly than the North Fork.

This basic structure was used to model the different policies, but the policies have not yet been fully explained. The following chapter will explain how the model is modified for the different policies, and give the results of the model.

NOTES

CHAPTER VI

1. The model below appeared in B. P. Baker, "Groundwater Protection Policy and Agricultural Production: A Recursive Stochastic Analysis," *Northeastern Journal of Agricultural and Resource Economics* 17(2) (October 1988): 131-138.

2. See Saul I. Gass, *Linear Programming*, (New York: McGraw-Hill, 1975).

3. Two sources which contain good reviews of the aggregation bias literature include Kenneth H. Baum, *A National Recursive Simulation and Linear Programming Model of Some Major Crops in U.S. Agriculture* (Ames, Iowa State University Ph.D. Dissertation, 1978): 82-98; and Brian W. Gould, *Energy Use, Environmental Quality, and Ethanol Production: An Analysis of The Impacts of Alternative Policies on Western New York* (Ithaca, Cornell University Ph.D. Dissertation, 1983): 29-33.

4. Richard H. Day, "Adaptive Economics and Natural Resources Policy," *American Journal of Agricultural Economics* 60 (1978): 276-283.

5. R. H. Day and A. Cigno, *Modelling Economic Change: A Recursive Programming Approach.*(Amsterdam: North-Holland Elsevier): 11.

6. J. W. Milon, "Interdependent Risk and Institutional Coordination for Nonpoint Externalities in Groundwater," *American Journal of Agricultural Economics* 68 (1986): 1229-1233.

7. S. S. Lazarus and G. B. White, "The Economic Potential of Crop Rotations in Long Island Potato Production," (Ithaca, NY: Cornell University Department of Agricultural Economics A.E, Res. 83-20, 1983); J. B. Phelps and R. B. How, "Planning Data for Small Scale Commercial Vegetable and Strawberry Production in New York," (Ithaca, NY: Cornell University Department of Agricultural Economics A.E, Res. 81-20, 1981); and Darwin Snyder, "Farm Cost Accounts," (Ithaca, NY: Cornell University Department of Agricultural Economics A.E, Res. 83-41, 1983).

8. Sources of labor statistics included the Riverhead office of the NY State Department of Labor; G. White, K. Gardner and D. Lee, "Survey of Farm Businesses in Suffolk County," *Suffolk County Agricultural News*, September 1983: 3-4; and the U.S. Census of Agriculture.

9. William Sanok, Suffolk County Farm Advisor, personal communication.

10. R. J. Wright, R. Loria, J. B. Sieczka and D. D. Moyer, "Final Report of the 1983 Long Island Potato Integrated Pest Management Program," (Cornell University, Department of Vegetable Crops, Mimeo 306, 1984).

11. The actual source code of the computer model, as well as other coefficients and data not included here are contained in Brian Baker, *Protection of Groundwater from Agricultural Pollutions: Institutions and Incentives.* (Ithaca, NY: Cornell University Ph.D. Dissertation, 1985).

Policy Analysis

Five alternative policies that deal directly with agricultural pollution are considered (Table 7.1). These policies are taxes and subsidies, control of cultural practices, pest control districts, changes in liability structure, and exclusion of agriculture from critical recharge areas. To analyze the effect of different policy on these two different objectives, it is necessary to understand the effect of policy on behavior that controls agriculture and the activities that pollute groundwater.

Table 7.1
Policies Considered for Analysis

---

1. Laissez Faire.

2. Ban aldicarb, carbofuran and oxamyl.

3. Tax on aldicarb, carbofuran and oxamyl.

4. Subsidized Conservation Crops.

5. Pest control districts.

All of the policies analyzed, with the exception of the base run and the policy taxing pesticides, assume that the pesticides which have been banned or removed from the market are not available for use on

Long Island. This assumption reflects that re-registration of the banned pesticides is politically unacceptable.

BAN ON SELECTED PESTICIDES

The first modification made of the laissez-faire model is a ban on aldicarb. This was done by simply adding a constraint to the LP model which required aldicarb loadings be less than or equal to zero.

The baseline model simulated farm production and pollution without a ban on pesticides. A constraint is introduced to reflect the elimination of the option to use banned pesticides. The baseline model is modified to analyze how different polices to protect groundwater change farm income and pesticide loading rates. The first is the ban on the pesticides aldicarb, carbofuran and oxamyl. Pest control decisions were based upon a choice between the pesticides kryocide, fenvalerate and rotenone.

The dual, or shadow price for the constraint banning aldicarb ranges between \$24 and \$47 per pound of active ingredient over the five years. This would be the amount a farmer should be willing to pay per pound of active ingredient of aldicarb if there were no constraints on the purchase of aldicarb. The banning of aldicarb results in its replacement by carbofuran, which dominates other pest control alternatives, but not to the degree that aldicarb did. Carbofuran, like aldicarb, was followed by replacement by substitute methods of pest control when resistance developed. However, the substitution is very small, as carbofuran is used on 100% of all potato land the first four years and 99% the fifth and final year.

The current policy, where aldicarb, carbofuran and oxamyl are not available to farmers for use, is the next considered. A constraint forcing carbofuran applications to be less than or equal to zero was added to the model (7.1).

$$(7.1) \qquad Z_\psi \quad \leq \quad 0$$

$$\psi = \text{aldicarb, carbofuran, oxamyl}$$

The oxamyl program was removed from the model, and a pest control decisions were based upon a choice between the pesticides kryocide, pydrin and rotenone.

TAXES AND SUBSIDIES

Economists have long favored taxes and subsidies as an efficient means for dealing with externalities.[1] A tax on externalities would make the private cost of those inputs more closely reflect the social cost they inflict.[2] A tax on agricultural inputs has several advantages over other measures to remedy non-point agricultural pollution. Because many producers are involved, the cost of enforcing practices would be high compared with the cost of setting and collecting a tax. A tax provides an incentive for the farmer to reduce the amount of the input. If the farmer reduces chemical inputs, the amount of pesticide or fertilizer which reaches the saturated zone of the soil is reduced.

Taxes and subsidies offer farmers the opportunity to re-allocate resources when correcting an externality. Several policy instruments are more likely to improve environmental quality and maintain production. A tax on agricultural inputs has several advantages over other measures to remedy pollution. The cost to enforce practices would be high compared with the cost to set and collect a tax when many producers are involved.[3]

Taxes on externalities have never received much attention in the United States outside of academic circles, despite their market efficiency.[4] The notable exception to aversion to this approach is Superfund, which deals primarily with hazardous waste. Potential problems with the tax arise from the enforcement and distortion of regional production. Any tax would probably have to be brought on at the Federal level. This is not only because of pre-emption. If pesticides were taxed by either the state or local government, farmers would simply go to another community to buy pesticides.

The efficient taxation of externalities present analysts with an infinite number of choices of how to model taxation. Optimal taxation requires information not only of the marginal product of the externality, but also of the social welfare trade-off between pollution and production.

Equation 7.2 gives the objective function for a policy to tax, rather than ban, pesticides found in the groundwater.

$$(7.2) \quad \text{MAX } E[\Pi_t] =$$

$$\sum_{i \in I} \sum_{k \in K} p_i y_{ik} - \sum_{i \in I} \sum_{k \in K} c_{ik} x_{ik} -$$

$$\sum_{\psi \in \Psi} \sum_{k \in K} (c_\psi \mp \Theta_\psi) - Z_{\psi k}$$

$$\{t=1, \ldots 5\}$$

Where Θ_ψ is the tax on pesticide ψ. Only the banned pesticides were taxed. Setting the optimal tax for pollution policy is empirically difficult.[5] If the tax is set too low, then the result is no different from the baseline. If the tax is set too high, the result is the same as a ban. Parametric programming was used to set the tax at a level that would eliminate the banned pesticides in an average year. If a CPB outbreak was significantly greater than average, the banned pesticides would enter the optimal solution. This would give farmers the option of applying materials during severe outbreaks that would otherwise be banned, but at a much higher price.

The value of the marginal product of pesticides can be estimated with some precision, but the preferences of society can only be reflected in the political process, as there is no market for pristine environment. The optimal tax on pesticide will vary from pesticide to pesticide, depending on its price, marginal product, toxicity, and environmental characteristics. A tax on all pesticides at the same rate would be an inefficient way of reducing potato acreage and may also fail to reduce the use of those which have the greater threat to the environment.

The next policy modelled was a subsidy for the cultivation of crops that required few, if any, agricultural chemicals. These are predominantly field crops with few pest, disease, or weed problems. Crops suitable for cultivation on Long Island that require few inputs and little labor include rye, oats, wheat, soybeans, sunflowers and dry beans. Other than rye, these do not have a local market outlet, and are less profitable than potatoes.[6] The subsidized objective function is:

$$(7.3) \qquad MAX\ E[\Pi_t] =$$

$$\sum_{i \in I} \sum_{k \in K} p_i y_{ik} - \sum_{i \in I} \sum_{k \in K} (c_{ik} + \sigma_i) x_{ik} -$$

$$\sum_{\psi \in \Psi} \sum_{k \in K} c_\psi z_{\psi k}$$

$$\{t = 1, \ldots 5\}$$

where σ_i is the subsidy for rye, oats, wheat, soybeans, sunflowers and dry beans, and zero for all other crops. A subsidy of $750 per acre was used. This level was sufficient to put 50% of all potato land into low-input crops. Annual payments ranged between $4 and $7 million. Subsidies lower than that would not offer farmers returns competitive with potatoes, and would therefore not be effective. Subsidies to achieve 100% participation would cost over $1,000 per acre. At that level, the subsidy would approach $20 million.

PEST CONTROL DISTRICTS

Integrated Pest Management (IPM) practices often involve externalities and public goods problems. If one farmer practices crop rotation as an IPM strategy, and his neighbors plant a host crop, that crop can serve as a reservoir for a mobile pest. If the host crop subsequently infests a rotated field, the beneficial effects of crop rotation are negated. A grower who releases beneficial insects cannot control their migration. If they migrate into a neighbors field, they will either provide the neighbor the economic benefit of reducing pest population, or the neighbor will spray and kill the beneficials. Either way, the farmer performing the release cannot capture all the benefits of biological control.

The market fails to provide IPM with the research necessary to develop alternatives to chemical methods. While a private market exists for the development of chemical control products, and for certain biological agents, the development of pest resistant varieties and many biological techniques cannot be completely captured by the market. Investment in basic research would very likely be suboptimal. Intervention in the market is justified where the public benefits of such research go beyond the private benefits. This has long been the justification for the land grant college system.

Districts have been created for other crops to manage pest and disease controls. Examples are citrus and cotton, districts in California.[7] If local control over pesticide application is to succeed, it must have approval from the state. A town ordinance restricted the use and transportation of pesticides, required the registration of applicators and banned three types of insecticides. This ordinance was struck down because either the state or the federal law had pre-empted the field.[8] Some states explicitly delegate regulatory authority over pesticides to local government.[9] A flexible system can give local needs and environmental concerns receive priority over state policy.[10]

States do have the power to delegate monitoring and enforcement of pesticide laws to the counties. California has created the authority for agricultural areas to form pest abatement districts.[11] These districts have the power to levy taxes,[12] purchase supplies, personal property, and real estate,[13] and is given broad powers to control pests as they are defined by the statute. The state court in California has ruled that this delegation of authority is constitutional.[14] A Federal court case has given California broad states rights in regulating pesticides.[15] A similar district could be formed for potatoes on Long Island.

Several tasks of the pest control district are simulated in the model. Specifically, these are that potatoes cannot be grown on the same field in two consecutive years. Pesticides are rotated to delay resistance and are selected to reduce environmental impacts to groundwater and wildlife. The district also bears the cost of scouting, and application of pest control materials. Participation is mandatory and region-wide. The crop rotation requirements of the pest control district are modelled by

$$(7.4) \qquad \sum_{k \in K_p} x_{kt} \leq 0.50 x_{k(t-1)}$$

where $x_{k(t-1)}$ is the potato acreage in period t-1 and x_{kt} is the acreage in potatoes in period t. Equation 7.4 places an upper limit on the acreage of potatoes following potatoes.

The expression in 7.5 prevents land from being in potatoes two years in a row. In the model, land in pest control districts cannot be cropped to potatoes for two consecutive years.

$$(7.5) \qquad \begin{cases} x_{kt} = 0 \ \text{ if } \ x_{k(t-1)} \in K_p \\[2em] x_{kt} \geq 0 \ \text{ if } \ x_{k(t-1)} \notin \overline{K}_p \end{cases}$$

Rotation from potatoes to grains are expected to disrupt the feeding of the CPB (Lazarus and White, 1984). Pesticide applications were limited to a synthetic pyrethroid with low mammalian toxicity and short half-life, fenvalerate, used with a synergist, PBO. A requirement was made that at least one fifth of the applications use rotenone to forestall the build-up of resistance. The policies described above were analyzed for their effect on farm income, environmental risk, and fiscal impact.

Support for pest control districts from farmers could be increased by subsidies. One form of subsidy is a purchase of chemical rights. Another way to provide an incentive for cooperation is for government to absorb all or most of the costs of administering and operating the districts.

PRODUCTION RESPONSE

The effect of the different policies on farm income, net of hired labor, variable costs, and purchased inputs, are presented in Table 7.2. The figures in parentheses represented government transfer payments. Discounting did not change the ranking of policies. Income loss associated with the ban on pesticides can be calculated to be approximately $2.2 million undiscounted over five years. The policy that results in the highest farm income is the conservation subsidy. This is followed by the baseline policy and the policy that taxes the pesticides that have been banned. The policies that ban the use of the carbamates aldicarb, carbofuran and oxamyl do not yield as high an income as the previously mentioned policies. The policy with the lowest income is the pest control district.

Table 7.2
Average Annual
Gross Margin Under Different Policies
-million dollars-

Rank	Policy	million $
	Conservation Subsidy	37.214
	(Subsidy Payment)	(5.931)
	Laissez Faire	35.145
	Tax on Pesticides	32.959
	(Tax Receipts)	(0.147)
	Ban on Pesticides	32.760
	Pest Control District	31.474

The conservation subsidy provides higher income, both through the subsidy itself, and through releasing labor tied up in potato production to the production of high value, labor intensive vegetable crops, such as cauliflower, cabbage, sweet corn, spinach and onions. The subsidy was set to attract one-third of the acreage in an average year. At this rate, the subsidy accounted for about 16% of farm income over a five year period. If the subsidy was too low, no farmer would adopt conservation practices. If the subsidy was set high enough, farmers would grow nothing but subsidized crops.

Farm income under the pest control district was lower than other policies, despite the subsidized services provided. The opportunity cost of restrictions on potato acreage more than offset the benefits of pesticide rotation and free scouting. The pest control district policy yields a farm income 4% less than the baseline and 15% less than the conservation subsidy.

Land use changes over the five year period are given in table 7.3. The change in crop acreage calculates the difference between 1988 predicted levels and 1984 actual levels in thousands of acres.

Potato acreage would continue to decline even if pesticides were not banned. Banning pesticides, however, accelerates the decline. Ironically, policies expressly designed to reduce potato acreage are less successful in acheiving that end than taxing potatoes.

Table 7.3
Predicted Changes in Crop Mix
1983-1988
(thousand acres)

Policy	Potatoes	Vegetables	Grain
Laissez Faire	-1.4	-0.2	-2.3
Ban on Pesticides	-2.4	-0.8	-1.5
Tax on Pesticides	-2.9	0.9	-1.7
Conservation Subsidy	-5.3	0.9	0.5
Pest Control District	-5.3	1.5	0.0

Grain includes the aggregate change in acreage of rye, wheat, field corn, soybeans, oats, and other grains. Grain acreage declines most rapidly under the laissez-faire scenario; a tax on potatoes causes grain acreage to increase the most. Bans on the pesticides slows the transition out of grain. Grain acreage increases with crop rotation, and stays about the same with the pest control district, purchase of crop rights and the ban on potatoes.

Vegetables are an aggregate of cauliflower, cabbage, sweet corn and other vegetables. For all scenarios but the laissez-faire, vegetable crop acreage increases. This is consistent with the move to specialty crops. The increase is most marked with the ban on potatoes and the tax on potatoes.

Too many factors are involved to confidently predict the fiscal impact of different policies. The cost of treating groundwater, or providing an alternative source of drinking water for wells that have already been contaminated is, in itself, a major undertaking. The most fiscally attractive policy is the taxation of pesticides, because it provides a source of revenue. Each alternative to the baseline policy has enforcement costs. While the baseline policy has no direct administrative costs, the great potential for leaching toxic chemicals makes its cost unpredictable and possibly great. The current policy of banning chemicals is attractive to regulatory agencies because of its predictability. Both the conservation subsidy and pest control district are potentially expensive.

ENVIRONMENTAL QUALITY INDICES

Economic effects alone would not adequately evaluate the different policies presented. Environmental risk is more difficult to quantify than farm income. Simulation of groundwater flows were notably unsuccessful at predicting groundwater contamination •ex ante•. Prediction of contamination levels over a large region would be inaccurate. Even if contamination levels could be accurately predicted, those levels would require heroic assumptions to translate them into dollar values.

Because of the multi-attribute nature of environmental hazard, a single number cannot give an absolute measure of risk. Pesticides have different characteristics and properties that make them have dissimilar environmental impacts. Solubility, adsorption, volatility, persistence and toxicity all contribute to the probability that a pesticide will leach into the water table and, if it does, the degree of threat to public health. Some are more toxic than others, some more persistent. Models of solute transport are based on the interaction of these physical, chemical and biological characteristics with soil, climate, management and crop cover. By a combination of these quantitative attributes, one can derive a relative measure of risk. The result is a index that can be used to rank alternative policies for their potential hazard to public health and the environment. Two indices are presented based on chemical and physical properties of different pesticides.

Indices used in the study were based on the physical, chemical and biological characteristics that influence the propensity of chemicals to leach into the water table and cause acute toxicity to consumers. The fate of pesticides depends on aqueous solubility (SOL_ψ), measured in mg/L; vapor pressure (V_y), measured in Pascals; and adsorption, (Koc_ψ), measured in L/Kg; and half-life ($t^1/_{2\psi}$), measured in days. Hydrogeology, soil and climate are also important in determining fate of pesticides. However, for the purpose of this study, these are assumed homogeneous for the region of study.

Leaching potential increases as half-life and solubility increase, and decreases as volatility (vapor pressure) and adsorption increases. As vapor pressure increases, more of the pesticide is volatilized and less is apt to reach the groundwater. Similarly, if a pesticide is likely to be adsorbed to soil particles, it is less likely to reach the groundwater. A leaching index based on these principles for each pesticide in the model ($LEACH_\psi$) is presented in 7.6.[16]

$$(7.6) \qquad \left[\quad LEACH_\psi \ = \ \frac{(SOL_\psi)(t^1/_{2\psi})}{(V_\psi)(Koc_\psi)} \quad \right]$$

The coefficients for the index are summarized in Appendix A.[17] Indices for individual pesticides are contained in table 7.3. The index is weighted by the pesticide loading rate of each pesticide in the model, Z_ψ, measured in pounds of active ingredients, by the characteristics for each pesticide, $LEACH_\psi$ and summed over all pesticides (7.7).

$$(7.7) \qquad HAZARD = \sum_{\psi \in \Psi} (LEACH_\psi)(Z_\psi)$$

Another index is constructed to take into account the acute toxicity of pesticide use under each policy. One measure of acute toxicity is LD50, the lethal dosage for half of a test animal population in a controlled experiment. As LD50 decreases, toxicity and therefore hazard increase. Leaching potential is divided by $LD50_\psi$, the LD50 for each pesticide. This index is used again to weight the pesticide loading rates from the model. This is represented by HAZARD (7.8).

$$(7.8) \quad LEACH = \sum_{\psi \in \Psi} \left[\frac{LEACH_\psi}{LD50_\psi} \right] (Z_\psi)$$

While the hazard index offers some measure of threat to human health, there are several precautions that should be taken when interpreting it. One cannot extrapolate the number of poisonings from the index, without making very strong assumptions about the functional form and relation between the characteristics in the index, as well as about the population at risk. One can only say that, in comparing two policies, one with a higher hazard index has a greater chance of causing acute toxicity incidents than another. The indices fail to account for chronic toxicity, such as cancer, and fail to discount for future risks to health. Synergistic effects are assumed to be nil, and the by-products of the decay of pesticides are assumed to be non-toxic. In spite of these limitations, the environmental indices here can be useful as guides for improving policy.

Table 7.4
Indices for Individual Pesticides

Generic Name	Trade Name	Leaching Index	Hazard Index
alachlor	Lasso	1.88E+06	1.04E+03
aldicarb	Temik	1.29E+12	1.39E+12
atrazine	Aatrex	3.66E+08	1.19E+05
captan	Captafol	2.05E+05	2.28E+01
carbaryl	Sevin	4.53E+03	9.06E+00
carbofuran	Furdan	3.70E+08	3.37E+07
DCPA	Dacthal	8.81E-08	2.94E-11
diazinon	Diazinon	1.55E+08	5.17E+05
endosulfan	Thiodan	1.12E-04	6.24E-06
EPTC	Eptam	8.81E+03	5.40E+00
fenvalerate	Pydrin	2.34E+11	5.18E+08
maneb	Manzate	1.58E-03	2.34E-07
parathion	Parathion	3.60E+03	9.00E+02

Source: Derived from equations 7.6 and 7.8, and the data in Appendix 1.

The leaching index gives an ordinal ranking of the likelihood that pesticides will reach the groundwater under different policies. The hazard index takes into account the acute toxicity of pesticides that are likely to contaminate groundwater. The results of the environmental indices are given in table 7.5.

The baseline policy has the highest leaching potential. This ranking reflects the high leaching potential of aldicarb used under this policy. The conservation subsidy has the lowest leaching potential, because it encourages the cultivation of crops that do not require the use of groundwater threatening pesticides.

There are several differences between the ranking of policies between the two indices. Rather than being the second most environmentally damaging policy as the leaching index indicates, pest control districts are the least damaging. This is brought about by the substitution of pesticides with low mammalian toxicity for more toxic ones used in the other policies.

Table 7.5
Environmental Indices

Policy	Leach Index	Hazard Index
Conservation subsidy	0.38	0.76
Ban on Pesticides	1.00	1.00
Tax on Pesticides	1.03	0.94
Pest Control District	1.44	0.60
Laissez Faire	59.53	11,539

The baseline policy is unquestionably the most harmful to human health and the environment. The evidence of this is indicated by the contamination levels of aldicarb and carbofuran in the drinking water near potato fields. Aldicarb would remain present in wells near fields in continuous potatoes. The banning of these pesticides is a sound remedy, and a good basis for other policies.

NOTES
CHAPTER VII

1. William J. Baumol and Wallace E. Oates *Theory of Environmental Policy*. (Englewood, NJ, Prentice-Hall, 1975): : 33-55; T. C. Schelling, *Incentives for Environmental Protection*, (Cambridge, MA: MIT Press, 1983).

2. A cogent analysis of the efficiency of taxing environmentally damaging activities can be found in Baumol and Oates: 152-190.

3. Gerald A. Carlson, "Economic Incentives for Pesticide Pollution Control" in J. Stephenson (ed) *The Practical Application of Economic Incentives to the Control of Pollution*. (Vancouver, University of British Columbia Press, 1974).

4. Frederick Anderson, *et al. Environmental Improvement Through Economic Incentive*. (Baltimore: Johns Hopkins Press, 1977).

5. Baumol and Oates, *A Theory of Externalities*.

6. Lazarus and White, The Economic Potential of Crop Rotations on Long Island.

7. See Brian Baker, "Pest Control in the Public Interest: Crop Protection in California" *UCLA Journal of Environmental Law and Policy* 8(1988): 31-71.

8. Long Island Pest Control Association v. Huntington, 341 N.Y.S.2d 93, (1973).

9. For example, Calif. Food and Agric. Code §14006.5 (West's Cum. Supp. 1988).

10. *People ex rel. Deukmejian v. County of Mendocino*, 204 Cal. Rptr. 897; 36 Cal. App. 3rd 476 (1984).

11. For citrus, Calif. Food and Agric. §§8401-8759 (West's Cum. Supp. 1988).

12. Calif. Food and Agric. §8551(d) (West's Cum. Supp. 1988).

13. Calif. Food and Agric. §8551(c) (West's Cum. Supp. 1988).

14. *Irvine v. Citrus Pest Control District No. 2 of San Bernardino County*, 144 P.2d 857, (1944).

15. *National Agr. Chemicals Ass'n v. Rominger*, 500 F.Supp. 465 (D.C. Cal. 1980).

16. D. A. Laskowski, C. A. I. Goring, P. J. McCall and R. L. Swann, "Terrestrial Environment, " in R. E. Conway (ed.) *Environmental Risk Analysis for Chemicals*. (New York: Van Nostrand, 1982).

17. Most $t^1/_2$, Koc, solubility, and vapor pressure coefficients were taken from Rao, Hornsby, and Jessup, (1985): 33; some of the solubility and vapor pressure coefficients were taken from Khan (1980): 218-223; *The Merck Index* (1983); *Pesticide Abstracts* (1983); Remaining Koc were either taken from Rao and Davidson (1980): 37-39; or were estimated using a regression from Laskowski, Goring, McCall, and Swann (1982): 212. LD_{50} figures were taken from New York State College of Agriculture and Life Sciences, *1984 New York State Pesticide Recommendations*: 13-20.

Conclusion

Research indicates that the ban on pesticides, while burdensome to farmers, still results in a positive return. Different policies to reduce pesticide use beyond banning them had little impact on the potential for groundwater contamination. Subsidies for low-input crops offer farmers the highest return of any policy, but at a cost for local government. Environmental quality indices suggest that water soluble carbamate pesticides ought not be used on Long Island. Other policies yielded indices close to one another, so the relative ranking may not be robust. However, the policy with the lowest potential for leaching pesticides is subsidy for conservation crops, while the one that poses the least hazard to public health is the formation of a pest control district.

One limitation of the model is that it does not adequately reflect transition to non-agricultural uses. Flexibility constraints control the pace at which farmland is retired, and is a naive part of the model. It is also limited because of the singular nature of Long Island.

Land use is complicated by the large and growing non-farm population in the region. Agricultural land preservation has cost local government millions of dollars, both through the purchase of development rights, and through property tax incentives. Policymakers have been reticent to impose any regulations that would further jeopardize Long Island agriculture. Polluting agricultural practices have undermined the public support that farmland preservation programs have received. The benefits afforded by open space are negated by groundwater contamination. Public policy on Long Island seeks the means to preserve farmland and preserve agriculture.

While Long Island is a unique situation, so is every contamination incident. There is no single optimal policy that can correct market failure in every case (Lipsey and Lancaster). This is particularly true of groundwater contamination where specific conditions and practices make generalizations misleading at best and impossible at worst. Long Island offers useful lessons because of the severity of the contamination incident.

AGRICULTURE IN TRANSITION

Potatoes have remained the dominant crop on Eastern Long Island throughout the 20th century. However, we have seen potatoes threatened by pests. Substantial acreage will remain in potatoes on Long Island for any policy save a moratorium on their growth. Potatoes will probably take up more acreage than any other crop for at least the near future. No other crop can provide as high a return given the labor constraints faced on the Island. The only crops which exceed potatoes in value are labor intensive fruits and vegetables. The regional availability of labor, particularly near harvest, would have to increase for fruit and truck crop production to expand. This would require a greater number of migrant workers, if permanent work is not available for hired farm workers.

Nevertheless, potatoes can be expected to decline in relative and absolute importance as policies, past, present and future, attempts to remedy the damage caused by pesticide contamination. Part of the acreage will be replaced by non-farm uses. The amount of land in agricultural will continue to decline, as long as the divergence between agricultural use value and market and development value remains high. Part will be replaced by orchard crops and other perennials.

As land is sold and developed, dimensions of remaining lands will be too small for the practical cultivation of potatoes. These plots will be about right for specialty crops which have a high cash value per acre. The mixed use of agricultural land with residential/second home development will enable subdividers to meet the new, stricter density requirements. While good data for this observation are lacking, there seems to be a trend toward diversification, one which should continue as the problems with potato production become more severe.

Potatoes will not disappear completely from the island. It is simply too profitable a crop, and the area is too well suited for their cultivation. One irony, however, is that the land with soils best suited for growing potatoes, Bridgehampton, is faced with the greatest pressure from non-agricultural uses.

ALTERNATIVE TECHNOLOGY

As the negative impacts of pesticide use become more apparent, farmers must develop alternative methods of insect, weed and disease control. Groundwater contamination is the not the only reason for developing these techniques, nor are pesticides the only source of agricultural contamination of groundwater. The situation is not unique to Suffolk County. Groundwater contamination is not limited to Long Island's potato farms. The National Academy of Sciences has reviewed alternative agricultural practices throughout the United States ahdn has determined that cultural, genetic and biological alternatives can greatly reduce pesticide use.[1]

In the case of potato production on Long Island, there are already several promising alternatives being developed. There are a variety of pathogens which attack the CPB. One with commercial potential as an agent for biological control is the fungus *Beauvaria bassiana*. The *Beauvaria* fungus attaches itself to the integument of the host, and grows hyphae which penetrate the insect's cuticle. The hyphae then invades and parasitized the host's circulatory and gastro-intestinal systems.[2] Toxins may also play a role in bringing about the death of the host organism.[3] Following the death of the host, hyphae penetrate the intersegmental membranes. Sporulation then occurs on the surface of the host's cadaver. *B. bassiana* is highly contagious.

A few predators and parasites of the CPB exist, but none have proven effective in naturally occurring populations at reducing CPB populations. Researchers have explored the introduction of non-native predators and parasites.[4] One potential parasite is the egg-parasite, *Edovum puttleri*. . When adults were released in infested potato fields, researchers reported that 90-95% of CPB eggs were destroyed.[5] If a low cost method to introduce the parasite can be developed, and if it can be a consistent egg parasite, *E. puttleri* can be an effective, integral part of CPB control. The breeding of a winter hardy race would do much to lower the cost of introduction.

Another development that shows promise at controlling CPB populations is the breeding of an arthropod resistant variety of potato. This is acheived by crossing wild species, such as *S. berthaultii* with *S. tuberosum*. These wild species have developed a resistance to arthropods. The mechanisms in these wild potato species which confer resistance to defoliation include foliar glykoalkaloids, glandular trichomes and proteinase inhibitors.[5] Wild potatoes were effective at inducing high mortality, slow larval growth, retarded development, reduced pupal and adult weights and low fecundity, compared with cultivated potatoes.[7]

The transition from an agriculture which depends on agricultural chemicals for pest control to a more integrated approach to pest management will undoubtedly cause economic hardship for some farmers. Many Suffolk County farmers are wary of government intervention. Nowhere in New York State will you find agriculture operating on such a large scale subject to as restrictive land use controls. Any program which threatens to further curtail the exercise of property rights will be viewed with suspicion. When confronting the issue of protecting groundwater from agricultural pollution, it is ironic to note that one argument used in the past to preserve agricultural land has been to protect groundwater quality.[8] The rationale behind this was that high density development causes increased saltwater intrusion, more septic leaching, household and lawn chemicals to intrude upon the groundwater. Agriculture is not the only alternative to high density development. Banning agriculture from critical recharge areas more or less runs counter to the goal of preserving agricultural land. The tradeoff between groundwater quality and the preservation of agriculture in Suffolk County presents a dilemma not easily solved.[9]

Diversification offers both economic and environmental benefits. When prices are volatile, farmers who diversify will suffer fewer losses than those who do not. Low potato prices may have as much, or more to do with farmers moving out of potato production as bans on pesticides. More crops means a more diverse ecosystem, one where insects, weeds and diseases do not become imbalanced.

Nitrate contamination was not examined in this study. Growers may have to consider rotations that minimize the need for soluble fertilizers, and use less soluble sources of nitrogen, such as legume cover crops and compost.[10] There are already encouraging signs that Long Island's agriculture is diversifying. The transition has not been unnoticed. This diversity needs to receive support. The transition will require new investment, and new capital. Perhaps the most important investment needed is in human capital, know-how. Structural change of this magnitude will cause, and is causing dislocation of many people. This period of adjustment is needed for the continuation of Long Island agriculture as a viable economic activity.

AGRICULTURE AND GROUNDWATER

Changes in agricultural technology have greatly increased the impact of food production on the environment. In making the tradeoff between private property and public good, the market favors immediate financial gain over long-run environmental quality. Intervention in the market to date has been reactive, rather than proactive. Chemicals have been banned only after they have been found in groundwater.

A more positive and forward-looking approach, beyond the scope of this study, is to foster an agriculture that does not depend on soluble nutrients and pesticides. The US Department of Agriculture, with its Low-Input Sustainable Agriculture (LISA) program, and several of the Land Grant Universities are pursuing this research. The amount of research performed on these problems is not sufficient to provide solutions. Even if funding were increased, it would be years before the alternatives could be developed and implemented. Moreover, as long as the cost of pesticide use is not adequately reflected in the price of pesticides, it will be cheaper to produce food that pollutes the environment than to grow food that internalizes externalities.

Further research is required on the economics of cultivating high value specialty crops with reduced inputs. Because of the lack of budgets for such cultivation practices, the conservation subsidy probably underestimates farm income and overestimates fiscal impact. Despite this, the conservation subsidy appears to be the most promising policy to protect groundwater quality and allow farms to remain solvent.

Pesticide contamination of groundwater involves more than the physical degradation of a resource; it is a problem of resource management with social, political and economic dimensions. There are tradeoffs between farm income, environmental quality, and fiscal responsibility. The choices made must reflect different sets of values, not all of which can be reflected in the market.

NOTES

CHAPTER VIII

1. National Research Council. *Alternative Agriculture.* (Washington, DC: National Academy Press, 1989): 175.

2. D. W. Roberts, R. A. LeBrun, and M. Semel. "Control of the Colorado Potato Beetle with Fungi," in J. H. Lashomb and R. Casagrande (eds.) *Advances in Potato Pest Management.* (Stroudsburg, PA: Hutchinson Ross Publishing Co., 1981): 119-137.

3. Sandra Galaini, *The Efficacy of Foliar Applications of Beauvaria bassiana Conidia Against Leptinotarsa decemlineata.* (Ithaca, NY: Cornell University Unpublished M.S. Thesis, 1984).

4. D. N. Ferro, "Pest Status and Control Strategies of the Colorado Potato Beetle," in D. N. Ferro and R. H. Voss (eds.) *Proceedings of the Symposium on the Colorado Potato Beetle.* (Amherst, MA: Massachussetts Experiment Station, 1985): 79-106.

5. R. F. W. Schroder and M. M. Athanas, "Review of Research on *Edovum puttleri Grissell,* Egg Parasite of the Colorado Potato Beetle," in D. N. Ferro and R. H. Voss (eds.) *Proceedings of the Symposium on the Colorado Potato Beetle.* (Amherst, MA: Massachussetts Experiment Station, 1985).

6. W. M. Tingey, "Potential for Plant Resistance in Management of Arthropod Pests," in J. H. Lashomb and R. Casagrande (eds.) *Advances in Potato Pest Management.* (Stroudsburg, PA: Hutchinson Ross Publishing Co., 1981): 268-86.

7. M. B. Dimock, *Studies on the Mechanisms of Resistance in the Wild Potato Solanum berthaultii (Hawkes) to the Colorado Potato Beetle, Leptinotarsa decemlineata (Say) (Coleoptera:*

Chrysomelidae). (Ithaca, NY: Cornell University Unpublished Ph. D. Dissertation, 1985).

8. Scholvinck, *Preserving Agriculture on Long Island*: 8-12.

9. An analysis of the economic and environmental tradeoffs can be found in Brian Baker, *Protection of Groundwater from Agricultural Pollution: Institutions and Incentives*. (Ithaca, Cornell University Ph.D. Dissertation, 1985); for a summary of the results of that study, and associated policy recommendations, see Brian Baker, "Can We Have Both Successful Farmers and Clean Groundwater on Long Island?" (Ithaca, Cornell University Dept. of Agr. Econ. Staff Paper, 1986).

10. Papendick, R. I., L. F. Elliot and James F. Power. (1987) "Alternative Production Systems to Reduce Nitrates in Groundwater." *American Journal of Alternative Agriculture* 2 (Winter): 19-24.

Bibliography

Adams, Walter and James W. Brock (1986) "Corporate Power and Economic Sabotage." *Journal of Economic Issues* 20: 919-940

Allee, David J. (1972) "The Interface of Public and Private Rights and Interests in Property." Ithaca: Cornell University Dept. of Agr. Econ. Staff Paper 72-28.

————. (1984) "Risk Assessment and Analysis". Ithaca: Cornell University Dept. of Agr. Econ. Staff Paper 84-4.

————. (1984) *Community in the Valley: A Challenge to the Capacity of Local Governance.* Ithaca: Cornell University Dept. of Agr. Econ. Staff Paper 84-10.

Allee, David J., et. al. (1977) "Integrating Water Quality and Water and Land Resources Planning." Staff Paper 77-26. Ithaca: Cornell University Dept. of Agr. Econ.

Alt, Klaus F. (1976) *An Economic Analysis of Field Crop Production, Insecticide Use and Soil Erosion in a Subbasin of the Iowa River.* Ames: Iowa State University Unpublished Ph.D. Dissertation.

Alt, Klaus F. and John A. Miranowski (1979) "Uncertainty and the Choice of Agricultural Abatement Policy." *Agric. Econ. Res.* 31(4) (October): 48-50.

Alt, Klaus F., John A. Miranowski and Earl O. Heady. (1979) "Social Cost and Effectiveness of Alternative Nonpoint Pollution

Control Practices." in Raymond C. Loehr, et. al. (eds.) *Best Management Practices*. Ann Arbor, MI: Ann Arbor Science Press.

Anderson, Frederick R., Allen V. Kneese, Phillip D. Reed, Russell B. Stevenson and Serge Taylor. (1977) *Environmental Improvement Through Economic Incentive*. Baltimore: Resources for the Future, Johns Hopkins Press.

Arthur, W. B. (1981) "The Economics of Risks to Life." *American Economic Review* 71(1): 54-64.

Ashton, Peter M. and Richard C. Underwood. (1975) *Non-Point Sources of Water Pollution*. Blacksburg: Virginia Polytechnic Institute and State University, Virginia Water Resources Center.

Baier, Joseph and Dennis Moran. (1981) *Status Report on Aldicarb Contamination of Groundwater as of September 1981*. Hauppauge, NY: Suffolk County Department of Health Services.

Baier, Joseph H. and Sy F. Robbins. (1982) Report of the Occurrence and Movement of Agricultural Chemicals in Groundwater: Suffolk County." Hauppauge, NY: Suffolk County Dept. of Health Services.

Baker, Brian P. (1985) *Protection of Groundwater from Agricultural Pollution: Institutions and Incentives*. Ithaca, NY: Unpublished Ph.D. dissertation.

————. (1988) "Groundwater Protection Policy and Agricultural Production: a Recursive Stochastic Analysis."

Northeastern Journal of Agricultural and Resource Economics 18(2): 131-138.

————. (1988) "Pest Control in the Public Interest: Crop Protection in California.") *UCLA Journal of Environmental Law and Policy* 8: 31-71

————. (1988) "Risk, Uncertainty and Technology Assessment," *Environment, Technology and Society.* 52: 8-10.

Baker, Emerson R. (1973) *Digest of State Pesticide Use and Application Laws.* Chamblee, GA: US EPA, Office of Pesticide Programs Operators Division

Baram, Michael S. (1980) "Cost-Benefit Analysis: An Inadequate Basis for Health, Safety, and Environmental Decisionmaking." *Ecology Law Quarterly* 8: 473-531.

Barker, R. and B. F. Stanton. (1965) "Estimating and Aggregation of Firm Supply Functions" *J. Farm Econ.* 47(3) (August): 701-712.

Barrows, Richard L. and Bruce A. Pengruber. (1975) "Transfer of Development Rights: An Analysis of a New Land Use Policy Tool" *Am. J. Agr. Econ.* 57(4) (November): 549-557.

Baum, Kenneth H. (1977) *A National Recursive Simulation and Linear Programming Model of Some Major Crops in U.S. Agriculture.* Ames: Iowa State University Unpublished Ph.D. Dissertation.

Baum, Kenneth H. and Lyle P. Schertz. (1983) *Modeling Farm Decisions for Policy Analysis.* Boulder: CO, Westview Press.

Baum, Kenneth H., et. al. (1981) "Econometric and Programming Tools for Evaluation of the Economic Impacts of Nonpoint Pollution" *Proceedings of Nonpoint Pollution Control Symposium* Rockville, MD: Interstate Commission on the Potomac River Basin. 274-284.

Baum, Kenneth H., Vincent Sposito and Earl O. Heady. (1979) "Recursive Interactive Programming: Computer Methodology and Procedures." Paper presented at the Seventh International Association of Agricultural Economists, Banff, Alberta, Canada, September.

Baumol, William J. and Wallace E. Oates. (1975) *The Theory of Environmental Policy.* Englewood Cliffs, NJ: Prentice-Hall.

Beltrami, Edward. (1982) *The High Cost of Clean Water*. Boston, Basel and Stuttgart: Birkhauser Press.

Black, Peter E. (1982) *Conservation of Water and Related Land Resources*. New York: Praeger Press.

Bloom, Sandra C. and Stanley E. Degler. (1969) *Pesticides and Pollution*. Washington, DC: Bureau of National Affairs.

Boisvert, Richard N. (1982) "Linear Programming." Ithaca: Cornell University Dept. of Agr. Econ.

Booth, Richard S. and Albert Bronson. (1983) *Major Institutional Arrangements Affecting Groundwater in New York State*. Ithaca: Cornell University Center for Environmental Research.

Bosso, Christopher. (1987) *Pesticides and Politics: The Life Cycle of a Public Issue*. Pittsburgh: University of Pittsburgh Press.

Boulding, Kenneth (1962) *A Reconstruction of Economics*. New York: Science Editions.

Boynton, Robert D. and Elizabeth A. Schiferl, (1983) *The Incidence of Rivalry in Pesticide Advertising*. Ithaca: Cornell University Dept. of Agr. Econ. Staff Paper 83-14.

Brady, Nyle C. (1974) *The Nature and Property of Soils*. New York: MacMillan Press.

Brady, Nyle C. (ed.) (1967) *Agriculture and the Quality of the Environment*. Washington, DC: American Association for the Advancement of Science.

Brickman, Ronald, Sheila Jasanoff and Thomas Ilgen. (1985) *Controlling Chemicals: The Politics of Regulation in Europe and the United States*. Ithaca, NY: Cornell University Press.

Brown, A. W. A. (1975) "Plant Protection from Pests" in A. W. A. Brown, et. al. (eds.) *Crop Productivity -- Research Imperatives*. E. Lansing, MI and Yellow Springs OH: Michigan Ag. Expt. Station and the Charles F. Kettering Foundation.

————. (1977) *Pesticide Management and Insecticide Resistance*. New York: Academic Press.

————. (1978) *The Ecology of Pesticides*. New York: Wiley and Sons.

Buchel, K. H. (1983) "Political, Economic and Philosophical Aspects of Pesticide Use for Human Welfare." in J. Miyamoto and P. C. Kearny (eds.) *Pesticide Chemistry: Human Welfare and the Environment*. New York: Pergamon Press.

Bunn, Dave. (1990) The Case Against Aldicarb. Los Angeles, CA: Pesticide Watch.

Burke, Denis P. (1977) "Common Scents: An Analysis of the Law of Feedlot Odor Control" *Creighton L. Rev.* 10: 539-591

California Assembly Office of Research (1985) *The Leaching Fields: A Nonpoint Threat to Groundwater*. Sacramento: California State Assembly.

Carlson, G. A. (1970) "A Decision Theoretic Approach to Crop Disease Prediction and Control." *Am. J. Agr. Econ.* 52(2) (May): 216-225.

————. (1974) "Economic Incentives for Pesticide Pollution Control" in J. Stephenson, (ed.) *The Practical Application of Economic Incentives to the Control of Pollution*. Vancouver, Univ. of British Columbia Press.

————. (1977) "The Long-Run Productivity of Insecticides." *Am. J. Ag. Econ.* 59(3) (August): 543-548.

————. (1984) "Risk Reducing Inputs Related to Agricultural Pests" in *Risk Analysis for Agricultural Production Firms: Concepts,*

Information Requirements and Policy Issues: 164-175. New Orleans, Conference Proceedings.

Caro, J. H. (1976) "Pesticides in Agricultural Runoff" in B. A. Stewart (ed.), *Control of Water Pollution from Cropland*: 91-119. Washington, US Environmental Protection Agency and US Department of Agriculture.

Caro, Robert A. (1974) *The Power Broker*. New York: Random House.

Carson, Rachel (1962) *Silent Spring*. Boston: Houghton Mifflin Co.

Cavelieri, Liebe. (1981) *The Double-Edged Helix*. New York, Columbia University Press.

Cetas, R. C., M. Semel and G. W. Selleck. (1978) Integrated Pest Management in Potatoes, 1978. Riverhead: Long Island Horticultural Research Laboratory Research Report.

Chavas, Jean-Paul, Rich Bishop and Kathy Segerson, (1986) "*Ex Ante* Consumer Welfare Evaluation in Cost-Benefit Analysis." *Journal of Environmental Economics and Management* 12: 255-268.

Clark, Robert M. and Richard G. Stevie. (1978) "Meeting Drinking Water Standards: The Price of Regulation." *Safe Drinking Water: Current and Future Problems*. Baltimore: Resources for the Future, Johns Hopkins University Press.

Clawson, Marion (1971) *Suburban Land Conversion in the United States: An Economic and Governmental Process*. Baltimore: Johns Hopkins Press for Resources For the Future.

Clement, Roland C. (1968) "The Pesticide Problem" *Nat. Res. J.* 8(1): 11-22.

Cohalan, Peter F. (1983) "Annual Environmental Report to the Suffolk County Legislature." Hauppauge, NY: Suffolk County Office of the County Executive.

Comment. "Groundwater Pollution in South Dakota: A Survey of Federal and State Law." *S. D. L. Rev.* 23: 698-734.

Comment. (1969) "Legal Control of Water Pollution" *U. Cal. Davis L. Rev.* 1: 141.

Comment. (1974) "Draft Proposal for Legislation to Control Water Pollution from Agricultural Sources." *Cornell L. Rev.* 59: 1097.

Comment. (1979) "Nonpoint Pollution Control in Virginia." *U. of Richmond L. Rev.* 13(3) (Spring): 539-556.

Congressional Research Service. (1984) *Groundwater Contamination by Toxic Substances: A Digest of Reports.* Washington, DC: US Library of Congress.

Conklin, Howard and William Lesher (1977) "Farm Value Assessment as a Means for Reducing Premature and Excessive Agricultural Disinvestment in the Urban Fringe." *Am. J. Ag. Econ.* 59(4) (November): 755-759.

Conrad, Jon. (1986) "On the Evaluation of Government Programs to Reduce Environmental Risk." *American Journal of Agricultural Economics* 68: 1272-1275.

Conway, Richard E. (ed.) (1982) *Environmental Risk Analysis for Chemicals.* New York: Van Nostrand Reinhold.

Copeland, Claudia (1980) *Nonpoint Pollution and the Area-Wide Waste Treatment Management Program Under the Federal Water Pollution Control Act.* Washington: Congressional Research Service, Library of Congress.

Coppock, Rob. (1984) "Control of Chemical Hazards: Risk Analysis, Conceptual Outlooks and Political Traditions in Selected Countries." *Journal of Public and International Affairs* 5: 67-83.

Corker, Charles E. (1971) *Groundwater Law, Management and Administration* Springfield, VA: National Technical Information Service.

Cornell University, Department of Rural Sociology. (1981) *National Statistical Assessment of Rural Water Conditions.* Ithaca: Cornell University.

Coye, Molly (1986) "The Health Effects of Agricultural Production," in Kenneth Dahlberg, *New Directions for Agriculture and Agricultural Research*

Crandall, Robert W. and Lester B. Lave (1981) *The Scientific Basis of Health and Safety Regulation.* Washington, DC: The Brookings Institute.

Cyert, Richard M. and James G. March (1963) *A Behavioral Theory of the Firm* Englewood Cliffs, NJ: Prentice Hall.

Davidson, J. M., P. S. C. Rao, L. T. Ou, W. B. Wheeler and D. F. Rothwell. (1980) *Adsorption, Movement and Biological Degradation of Large Concentrations of Pesticides in Soils.* Springfield, VA: National Technical Information Service.

Davidson, J. M., P. S. C. Rao, L. T. Ou, W. B. Wheeler and D. F. Rothwell. (1980) *Adsorption, Movement and Biological Degradation of Large Concentrations of Selected Pesticides in Soil.* Springfield, VA: National Technical Information Service.

Davis, Peter N. (1971) "Theories of Water Pollution Litigation" *U. Wisc. L. Rev.* 1971(3): 738-816.

―――. (1976) "Groundwater Pollution: Case Law Theories for Relief" *Mo. L. Rev.* 39: 117-159

Day, Richard H. (1961) "Recursive Programming and Supply Prediction" in Earl O. Heady, *Agricultural Supply Functions* Ames: Iowa State University Press.

————. (1963) "On Aggregating Linear Programming Models of Production." *J. Farm Econ.* 45(3): 797-813.

————. (1963) *Recursive Programming and Production Response.* Amsterdam: North Holland Elsevier.

————. (1969) "More on the Aggregation Problem: Some Suggestions." *Am. J. Ag. Econ.* 51(3) (August): 686-688.

————. (1978) "Adaptive Economics and Natural Resources Policy" *Am. J. Agr. Econ.* 60(2) (May): 276-283.

————. (1983) "Farm Decisions, Adaptive Economics and Complex Behavior in Agriculture" in K. H. Baum and L. P. Schertz (eds.) *Modeling Farm Decisions for Policy Analysis* Boulder, CO: Westview Press.

Day, Richard H. and Alessandro Cigno (eds.) (1978) *Modelling Economic Change: The Recursive Programming Approach* Amsterdam: North Holland Elsevier.

Dean, Gerald W. and Michele DeBenedictis (1964) "A Model of Economic Development for Peasant Farms in Southern Italy" *J. Farm Econ.* 46(2): 295-312.

DeBach, Paul (1974) *Biological Control by Natural Enemies.* New York: Cambridge University Press.

de la Mora, Gabriella. (1985) *Long Island Potato Markets and Markets for Alternative Vegetable Crops.* Ithaca, NY: Cornell University Unpublished M.S. Thesis

Delfino, Joseph J. (1977) "Contamination of Potable Groundwater Supplies in Rural Areas" in Robert B. Pojacek (ed.) *Drinking Water Quality Enhancement Through Source Protection*: 275-295. Ann Arbor: Ann Arbor Science Press.

Dethier, B. E. and M. T. Vittum. (1967) *Growing Degree Days in New York State*. Ithaca: Cornell University Agricultural Experiment Station.

Dimock, M. B. (1985) *Studies on the Mechanisms of Resistance in the Wild Potato Solanum berthaultii (Hawkes) to the Colorado Potato Beetle Leptinotarsa decemlineata (Say) (Coleoptera: Chrysomelidae)*. Ithaca, NY: Cornell University Unpublished Ph. D. Dissertation.

Dixon, Orani, Peter Dixon and John Miranowski. (1973) "Insecticide Requirements of an Efficient Agricultural Sector. *Rev. Econ. Stats.* 55(4) (November): 423-432.

Dorfman, Robert. (1980) "Environmental Responsibilities of Agricultural Economics and Operations Research." in D. Yaron and C. Tapiero (eds.) *Operations Research in Agriculture and Water Resources*: 3-16. Amsterdam: North-Holland Publishing Company.

Downing, P. B. (1981) "A Political Economy Model of Implementing Pollution Laws." *J. Environ. Econ. Mgmt.* 8(3): 255-271.

Downs, Anthony. (1972) "Up and Down with Ecology--The 'Issue-Attention Cycle." *Public Interest* 28: 38-50.

Dworsky, Leonard B. and David J. Allee. (1979) *The Integration of Water Quantity and Water Quality Management*. Staff Paper 79-1. Ithaca: Cornell University Dept. of Agr. Econ.

Edmunds, Stahrl W. (1980) "Environmental Policy: Bounded Rationality Applied to Unbounded Ecological Problems." *Policy Studies Journal* 3(9): 359-369.

Edwards, W. F. (1969) *Economic Externalities in the Agricultural Use of Pesticides and the Evaluation of Alternative Policies*. Gainesville: Univ. of Florida Unpublished Ph.D. Dissertation.

Ellickson, Robert C. and A. Dan Tarlock. (1981) *Land Use Controls: Cases and Materials*. Boston: Little, Brown and Company.

Environmental Law Institute. (1977) *Legal and Institutional Approaches to Water Quality Management Planning and Implementation*. Washington DC: US Government Printing Office.

ERM-Northeast and Camp Dresser & McKee (1983) *North Fork Water Supply Plan Suffolk County, New York*. Hauppauge, NY: Suffolk County Department of Health Services.

Ezrahi, Yaron. (1983) "Utopian and Pragmatic Rationalism: The Political Context of Scientific Advice." *Minerva* 18: 111-131.

Fairchild, Deborah M. (1987) "A National Assessment of Ground Water Contamination from Pesticides and Fertilizers," in Fairchild (ed.) *Ground Water Quality and Agricultural Practices: 273-294*. Chelsea, MI: Lewis Publishers.

Ferro, D. N. et. al. (1983) "Crop Loss Assessment of the Colorado Potato Beetle on Potatoes in Western Massachussetts." *J. Econ. Entomology*. 76:349-356 (April).

————. (1985) "Pest Status and Control Strategies of the Colorado Potato Beetle. In Proceedings of the Symposium on the Colorado Potato Beetle, XVIIth International Congress of Entomology, D. N. Ferro and R. H. Voss (eds.) Massachussetts Experiment Station, Amherst.

Fertig, Stanford. (1978) "The RPAR Process and the Impact of Regulatory Decisions on Agriculture." *Proceedings of the Southern Weed Science Society* 34: 6-16..

Fessenden-Raden, June and Rita R. Calvo. (1986) "Aldicarb in Groundwater: Determining Acceptable Risk." *General Ecology's Water Research Update* 2(2).

Flint, Mary Lou and Robert van den Bosch, (1981) *Introduction to Integrated Pest Management*. New York: Plenum Press.

Fohner, G. R. (1984) Long Island Vegetable and Potato Rotations. Ithaca: Cornell University Dept. of Agr. Econ.

Fohner, G. R. and G. B. White. (1981) *Pesticide Use on Potatoes in Upstate New York*. A. E. Res. 81-7. Ithaca: Cornell University Dept. of Agr. Econ.

————. (1982) *Cost of Pesticides for Potatoes in Upstate New York, 1981*. A. E. Res. 82-30. Ithaca: Cornell University Dept. of Agr. Econ.

Forgash, Andrew J. (1981) "Insecticide Resistance of the Colorado Potato Beetle, Leptinotarsa decemlineata (Say)." in Lashomb & Casagrande (eds.) *Advances in Potato Pest Management*. Stroudsburg, PA: Hutchinson Ross Publishing Co.

————. (1985) Insecticide resistance in the Colorado potato beetle. In *Proceedings of the Symposium on the Colorado Potato Beetle, XVIIth International Congress of Entomology*, D. N. Ferro and R. H. Voss (eds.) Massachussetts Experiment Station, Amherst.

Forrester, Jay W. (1961) *Industrial Dynamics*. Cambridge: MIT Press.

Fox, Austin S. (1971) "The Economic Consequences of Restricting or Banning the Use of Pesticides." in Ronald L. Mighell (ed.), *Economic Research on Pesticides for Policy Making*. Washington, DC: US Department of Agriculture, Economic Research Service.

Francis, J. et. al. (1981) *National Statistical Assessment of Rural Water Conditions*. Ithaca: Cornell University Department of Rural Sociology.

Freeman, A. Myrick III. (1979) *The Benefits of Environmental Improvement Theory and Practice*. Baltimore: Resources for the Future, Johns Hopkins University Press.

————. (1982) "Risk Evaluation in Environmental Regulation, in Wesley A. Magat (ed.) *Reform of Environmental Regulation*, Cambridge, MA: Ballinger Publishing.

Freeman, A. Myrick III and Robert Haveman. (1972) "Clean Rhetoric and Dirty Water." *Public Interest* 28.

Freeze, R. Allan and John A. Cherry. (1975) *Groundwater.* Englewood Cliffs, NJ: Prentice-Hall.

Galaini, Sandra. (1984) *The Efficacy of Foliar Applications of Beauvaria bassiana Conidia Against Leptinotarsa decemlineata.* Ithaca, NY: Cornell University unpublished Ph. D. dissertation.

Galbraith, John Kenneth. (1952) *American Capitalism: The Concept of Countervailing Power.* Boston: Houghton-Mifflin.

Gardner, Kenneth V. (1983) "Agricultural District Legislation in New York." A. E. Ext. 83-17. Ithaca: Cornell University Dept. of Agr. Econ.

Gass, Saul I. *(1975) Linear Programming.* New York: McGraw-Hill.

Gauthier, Norman L. Richard N. Hofmaster and Maurie Semel. (1981) "History of Colorado Potato Beetle Control" in Lashomb & Casagrande (eds.) *Advances in Potato Pest Management.* Stroudsburg, PA: Hutchinson Ross Publishing Co.

Gelpe, Marcia. (1981) "Animal Feedlot Regulation in Minnesota." *William Mitchell L. Rev.* 7(2): 399-457.

Georgescu-Roegen, Nicholas. (1971) *The Entropy Law and the Economic Process.* Cambridge: Harvard University Press.

Georghiu, George P. (1986) "The Magnitude of the Resistance Problem," in *Pesticide Resistance: Strategies and Tactics for Management of Pesticide Resistant Pest Populations.* Washington, DC: National Academy Press.

Ghatak, Subrata, Kerry Turner and Anita Ghatak. (1981) "Benefits and Costs of Pesticide Use and Some Policy Implications for Less Developed Countries." in Manas Chatterji (ed.) *Energy and Environment in the Developing Countries*. New York: John Wiley and Sons.

Goldberger, Paul. (1983) "The Strangling of a Resort." *New York Times Magazine*. September 4: 14 ff.

Gould, Brian W. (1983) *Energy Use, Environmental Quality and Ethanol Production: An Analysis of the Impacts of Alternative Policies on Western New York*. Ithaca: Cornell University Unpublished Ph. D. Dissertation.

Graham, Frank Jr. (1970) *Since Silent Spring*. Boston: Houghton Mifflin Co.

Graham, John D. and James W. Vaupel. (1981) "Value of a Life: What Difference Does it Make?" *Risk Analysis* 1(1): 89-95.

Gramlich, Edward M. (1981) *Benefit-Cost Analysis of Government Programs*. Englewood Cliffs, NJ: Prentice Hall.

Greenburg, Ellen, Sarah Meyland and James T. B. Tripp (eds.) (1982) "Watershed Planning for the Protection of Long Island's Groundwater." Coalition for the Protection of Long Island's Groundwater.

Griffin, Ronald C. and Daniel W. Bromley. (1982) "Agricultural Runoff as a Nonpoint Externality: A Theoretical Development." *Am. J. Agr. Econ.* 64(3): 547-552 (August).

Group for the South Fork. (1982) *Groundwater Management: A Handbook for the South Fork*. Bridgehampton, NY: Group for the South Fork, Inc.

Hadley, George (1964) *Non-Linear and Dynamic Programming*. Reading, MA: Addison-Wesley Publishing Co.

Haimes, Yacov Y. (1981) *Risk/Benefit Analysis in Water Resources Planning and Management*. New York: Plenum Press.

Haimes, Yacov Y. and David J. Allee (1984) *Multiobjective Analysis in Water Resources. New York: American Society of Civil Engineers.*

Hallberg, George R. (1986) "Overview of Agricultural Chemicals in Ground Water," *Agricultural Impacts on Ground Water--A Conference*. Dublin, OH: Water Well Journal Publishing Co.: 1-66.

Hamilton, A. and L. W. Libby (1978) "The Policy Relevance of Alternative Institutional Approaches to 208 Planning." in Ray C. Loehr (ed.) *Best Management Practices in Agriculture and Silviculture*. Ann Arbor, MI: Ann Arbor Science Press.

Hand, Jacqueline P. (1984) "Right to Farm Laws: Breaking New Ground in the Preservation of Farmland." *U. Pitt. L. Rev.* 45 (Winter): 289.

Hardy, Ernest E. (1969) *The Agricultural Regions of New York State. Ithaca, NY:* Cornell University unpublished Ph. D. dissertation.

Hare, J. Daniel. (1980) "Impact of Defoliation by the Colorado Potato Beetle in Potato Yields." *J. Economic Entomology*. 73(3) (June): 369-373.

Headley, J. C. (1968) "Estimating the Productivity of Agricultural Pesticides." *Am. J. Ag. Econ.* 50(1) (February): 13-23.

Headley, J. C. and J. N. Lewis. (1966) *The Pesticide Problem: An Economic Approach*. Baltimore: Resources for the Future, Johns Hopkins University Press.

Henderson, Hazel. (1978) *Creating Alternative Futures*. New York: G. P. Putnam.

Henderson, James M. (1959) "The Utilization of Agricultural Land: A Theoretical and Empirical Inquiry" *Rev. Econ. Stat.* 41(3): 242-259.

Henderson, T. R., J. Traubman, and T. Gallagher. (1985) *Groundwater: Strategies for State Action.* Washington, DC: The Environmental Law Institute.

Hennigan, Robert D. (1981) Water Supply Source Protection Rules and Regulations Project: Final Report. Syracuse: N.Y. State College of Environmental Science and Forestry.

Herman, Lydia, Richard Carlson, Theodore Goldfarb and Barry Commoner. (1982) The New York Metropolitan Area Produce Market: A New Opportunity to Preserve Long Island Farmland. Flushing, NY: Center for Biology of Natural Systems, Queens College.

Hillebrandt, P. M. (1960) "The Economic Theory of the Use of Pesticides" *J. Agr. Econ.* 13(4) (January): 464-472.

Hines, N. William (1974) "Farmers, Feedlots and Federalism: The Impact of the 1972 Federal Water Pollution and Control Act Amendments on Agriculture." *S.D. L. Rev.* 19 (Summer): 540-566.

Hines, N. William. (1970) "Agriculture: The Unseen Foe in the War on Pollution." *Cornell L. Rev.* 55: 740-760.

Holmes, Beatrice H. (1979) *Institutional Bases for Control of Nonpoint Source Pollution.* Washington, DC: US EPA, Office of Water and Waste Management.

Hueth, D. and U. Regev. (1974) "Optimal Agricultural Pest Control with Increasing Pest Resistance." *Am. J. Agr. Econ.* 56(3) (August): 543-552.

Huffaker, Jeffrey D. (1978) "Regulation of Pesticide Use in California." *U.C. Davis L. Rev.* 11(2) (November): 273-299.

Hughes, Henry B. F. and Keith Porter. (1984) "Interim Results Tracking Aldicarb Residues in Long Island Groundwater." Ithaca: Cornell University Center for Environmental Research.

I.B.M. (1971) Mathematical Programming System-Extended (MPSX) Control Language User's Manual. White Plains, NY: IBM.

Irvine, David E. G. and Brian Knights. (1974) Pollution and the Use of Chemicals in Agriculture. London: Butterworths.

Jansson, Erik. (1979) "Field Reports on Integrated Pest Management and Biological Control: Is the Farmer Making a Profit from These Programs, and How Much?" Agricultural and Environmental Relationships: Issues and Priorities. Washington, DC: Congressional Research Service.

Jenkins, S. H. (ed.) (1979) *The Agricultural Industry and Its Effects on Water Quality.* New York: Pergamon Press.

Josephson, Julian. (1980) "Groundwater Strategies." *Environ. Sci. & Tech.* 14(9) (September): 1030-1035.

Junger, Peter. (1976) "Bad Water: Common Law Nuisance and Welfare Economics Mixed Well". *Case West. L. Rev.* 27: 3.

Jury, W. A., W. F. Spencer, and W. J. Farmer. (1983) "Behavior Assessment Model for Trace Organics in Soil: I. Description of Model." *J. Environ. Qual.* 12: 558-564.

Just, Richard and David Zilberman. (1979) "Asymmetry of Taxes and Subsidies in Regulating a Stochastic Mishap." *Q. J. Econ.* 93(1) (February): 139-148.

Kahneman, Daniel, Paul Slovic and Amos Tversky. (1982) *Judgement under Uncertainty: Heuristics and Biases.* Cambridge: Cambridge University Press.

Kalt Joseph P. and Mark A. Zupan. (1985) "Capture and Ideology in the Economic Theory of Politics." *American Economic Review* 74: 279-300.

Kelman, Stephen J. (1981) "Cost-Benefit Analysis: An Ethical Critique" *Regulation* 36.

Kenaga, Eugene E. (1977) "Evaluation of the Hazard of Pesticide Residues in the Environment" in Watson & Brown (eds.) Pesticide Management and Insecticide Resistance: 51-95. New York: Academic Press.

————. (1982) "Evaluation of Pesticide Impacts on the Environment-- Shortcut Methods for Hazard Assessment" Proceedings of a Workshop on Agrichemicals and Estuarine Productivity: 177-213. Beaufort, NC: Duke University Marine Laboratory.

Klein, John V. N. (1973) "Farmlands Preservation Program Report to the Suffolk County Legislature." Hauppauge, NY: Suffolk County Office of the County Executive.

Kneese, Allen V. (1984) *Measuring the Benefits of Clean Air and Water*. Baltimore: Johns Hopkins Press.

Kneese, Allen V. and Blair T. Bower. (1979) *Environmental Quality and Residuals Management*. Baltimore: Resources for the Future, Johns Hopkins Press.

Kneese, Allen V. and Charles L. Schultze. (1975) *Pollution, Prices and Public Policy*. Washington: Brookings Institute.

Koppelman, Lee. (1984) Status Sheet of the Suffolk County Farmland Preservation Program." Hauppauge, NY: Long Island Regional Planning Board.

LaDue, Eddy L., Christine Shoemaker, Noel P. Russell, Robert B. Rovinsky and David Pimentel. (1979) "The Potential Impact of Cotton Insect Control Technology." Cornell Univ. Agr. Econ. Staff Paper 79-31.

Lake, Elizabeth E., William N. Hanneman and Sharon M. Oster. (1979) *Who Pays for Clean Water*. Boulder, CO: Westview Press.

Lansky, David M. (1984) *Phenology of the Colorado Potato Beetle Population on Long Island*. Ithaca, NY: Cornell University unpublished Ph. D. dissertation.

Lashomb, James H. and Richard Casagrande. (1981) *Advances in Potato Pest Management*. Stroudsburg, PA: Hutchinson Ross Publishing Co.

Laskowski, Dennis A., Cleve A. I. Goring, P. J. McCall, and R. L. Swann. (1982) "Terrestial Environment." in R. E. Conway (ed.) *Environmental Risk Analysis for Chemicals*. New York: Van Nostrand Reinhold.

Lave, Lester (1981) *The Strategy of Social Regulation*. Washington, DC: Brookings Institute.

————. (1982) *Quantitative Risk Assessment in Regulation*. Washington: Brookings Institute.

Lazarus, S. S. and G. B. White. (1983) The Economic Potential of Crop Rotations in Long Island Potato Production. Cornell University Agr. Econ. Res. 83-20.

Lesher, William G. and Doyle A. Eiler. (1978) "An Assessment of Suffolk County's Farmland Preservation Program." *Am. J. Agr. Econ.* 60(1) (February): 140-143.

Lichtenberg Erik, and David Zilberman (1985) Efficient Regulation of Environmental Health Risks. San Francisco, Western Consortium for the Health Professions

————. (1986) "Problems of Pesticide Regulation: Health and Environment versus Food and Fiber," in Tim T. Phipps, Pierre R. Crosson and Kent A. Price (eds.) *Agriculture and the Environment:* : 123-145. Washington, Resources for the Future.

Linnerooth, Joanne (1979) "The Value of Human Life: A Review of the Models." *Economic Inquiry* 17: 52-74.

Logan, Patrick A. (1981) "Estimating and Projecting Colorado Potato Beetle Density and Potato Yield Loss." in Lashomb and Casagrande (eds.) *Advances in Potato Pest Management.* Stroudsburg, PA: Hutchinson Ross Publishing Co.

Logan, Patrick A. and Richard A. Casagrande. (1980) "Predicting Colorado Potato Beetle (Leptinotarsa decemlineata Say) Density and Potato Yield Loss." Environmental Entomology 9(5) (October): 659-663.

Long, Larry and Diane DeAre. (1982) "Repopulating the Countryside: A 1980 Census Trend". *Science* 217: 1111-1115.

Long Island Regional Planning Board (1984) "Status Sheet of the Suffolk County Farmland Preservation Program." Hauppauge, NY. Lord, William B.

Lowi, Theodore. (1969) *The End of Liberalism.* New York: W. W. Norton.

Lowrance, W. W. (1976) *Of Acceptable Risk.* Los Altos, CA: Kaufmann.

Magnuson, Paula A. (1983) "Uses of Different Types of Aquifers." *Groundwater Resources and Contamination in the United States*: 151-172. Washington, DC: National Science Foundation.

Marshall, Eliot. (1983) "The Murky World of Toxicity Testing." *Science* 220: 1130-1131.

————. (1985) "The Rise and Decline of Temik," *Science* 229: 1369-1371.

Massey, Dean T. (1983) "Land Use Regulatory Power of Conservation Districts in the Midwestern States for Controlling Nonpoint Source Pollutants." *Drake L. Rev.* 33: 35-111.

May, Linda Lorraine. (1984) *Dynamic Risk Benefit Analysis in Pesticide Regulation: The Case of Toxaphene.* Riverside: University of California Unpublished Ph.D. Dissertation.

McGrann, J. M. and J. Meyer. (1978) "Farm Level Economic Evaluation of Erosion Control and Reduced Chemical Use in Iowa." in Raymond C. Loehr, et. al. (eds.) *Best Management Practices for Agriculture and Silviculture.* Ann Arbor, MI: Ann Arbor Science Press.

McKean, Roland N. (1980) "Enforcement Costs in Environmental and Safety Regulations." *Policy Analysis* 6: 269-289.

Meyers, Charles J. and A. Dan Tarlock. (1980) *Water Resource Management* (2d Ed.) Mineola, NY: Foundation Press, Inc.

Midwest Research Institute and RvR Consultants. (*A Study of the Efficiency of the Use of Pesticides in Agriculture.*) Springfield, VA: National Technical Information Service.

Mighell, Ronald L. (ed.) (1971) *Economic Research on Pesticides for Policy Decisionmaking.* Washington, DC: US Department of Agriculture, Economic Research Service.

Miller, Lowell E. (1977) "Pesticide Residue Regulation." in Watson and Brown (eds.) *Pesticide Management and Insecticide Resistance:* 129-135. New York: Academic Press.

Milon, J. W. (1986) "Interdependent Risk and Institutional Coordination for Nonpoint Externalities in Groundwater," *American Journal of Agricultural Economics* 68: 1229-1233.

Miranowski, John A. (1975) *Crop Insurance and Information Services to Control Use of Pesticides.* Springfield, VA: National Technical Information Service.

————. (1975) *The Demand for Agricultural Crop Chemicals Under Alternative Farm Programs and Pollution Control Solutions."* Cambridge, MA: Harvard Univ. Unpublished Ph. D. Dissertation.

Miranowski, John A., Ulrich F. W. Ernst and Francis H. Cummings. (1974) *An Investigation of Federally Sponsored Crop Insurance and Information Services to Control Use of Pesticides.* Cambridge, MA: ABT Associates Inc.

Mishan, Ezra. (1978) *Cost-Benefit Analysis.* New York: Praeger

Moffit, L. J. and R. L. Farnsworth. (1981) "Bioeconomic Analysis of Pesticide Demand." *Agric. Econ. Res.* 33(4):12-18.

Montgomery, John E. (1976) "Control of Agricultural Water Pollution: A Continuing Regulatory Dilemma." *Univ. Ill. Law Forum.* 1976: 533-559

Mumford, J. D. and G. A. Norton. (1984) "Economics of Decision Making in Pest Management." *Annual Review of Entomology.* 29: 157-174.

Muto, Kazuo. (1964) *Predicting Acreage of Major Crops in New York State Using Recursive Programming.* Ithaca, NY: Cornell University unpublished Ph. D. Dissertation.

Muzzey, Susan E. (1983) "The Safe Drinking Water Act and the Realities of Groundwater Pollution." *St. Louis U. Law J.* 27: 1019-1038.

National Agricultural Lands Survey (1979) *Final Report.* Washington, DC: U.S. Government Printing Office.

National Research Council. (1977) *Decision Making in the Environmental Protection Agency.* Washington, DC: National Academy of Sciences.

————. (1980) *Regulating Pesticides.* Washington, DC: National Academy of Sciences.

————. (1983) *Risk Assessment in the Federal Government: Managing the Process.* Washington, DC: National Academy of Sciences.

————. (1987) *Regulating Pesticides in Food: The Delaney Paradox.* Washington, DC: National Academy Press.

————. (1989) *Alternative Agriculture.* Washington, DC: National Academy Press.

Neal, R. A. (1981) "Risk Benefit Analysis: The Role of Regulation in Pesticide Registration." in S. Kris Bandel, Gino J. Marco, Leon Golberg and Marguerite L. Leng (eds.) *The Pesticide Chemist and Modern Toxicology* Washington, American Chemical Society Symposium 160: 469-472.

Nelson, Richard R. and Sidney G. Winter. (1982) *An Evolutionary Theory of Economic Change.* Cambridge, MA: Belknap Press.

New York State College of Agriculture and Life Sciences. (1984) *1984 Cornell Recommends for Field Crops.* Ithaca: Cornell University.

————. (1984) *1984 Cornell Recommends for Vegetable Crops.* Ithaca: Cornell University.

————. (1984) *New York Pesticide Recommendations. (Red Book).* Ithaca: Cornell University.

New York State Department of Agriculture and Markets. (1982) *Agricultural Statistics* Albany: New York State.

New York State Department of Conservation. (1978) *Groundwater Classifications, Quality Standards and/or Limitations.* Albany: New York State.

Newton, David F. (1982) "Suffolk County's Farmland Preservation Program." Riverhead, Suffolk County Cooperative Extension.

Newton, D. and M. Boast. (1978) "Preservation by Contract: Public Purchase of Development Rights in Farmland." *Columbia Journal Environ. Law* 4: 190

Nielsen, Elizabeth G. and Linda K. Lee. (1987) "The Magnitude and Costs of Groundwater Contamination from Agricultural Chemicals: A National Perspective." Washington, DC: US Department of Agriculture, Economic Research Service.

Noble David F. and Nancy E. Pfund (1980) "Business Goes Back to College." *The Nation* 231 (September 20): 247-248.

Norgaard, Richard B. (1976) "Integrating Economics and Pest Management." in J. Lawrence Apple and Raymond F. Smith (eds.) *Integrated Pest Management*. New York: Plenum Press.

Norman, J. M. (1972) *Heueristic Procedures in Dynamic Programming*. Manchester, England: Manchester University Press.

Novotny, Vladimir and Gordon Chesters. (1981) *Handbook of Nonpoint Pollution Sources and Management*. New York: Van Nostrand Reinhold.

O'Brien, Mary. (1990) "If Not Risk Assessment, then What?" *Journal of Pesticide Reform* 10: 2-6.

Olsen, Mancur. (1965) *The Logic of Collective Action*. Cambridge, MA: Harvard University Press.

Osteen, Craig D., E. B. Bradley and L. Joe Moffitt. (1981) *The Economics of Agricultural Pest Control: An Annotated Bibliography 1960-1980*. Washington, DC: US Department of Agriculture.

Ou, L. T., J. M. Davidson and P. S. C. Rao. (1980) *Rate Constants for Transformation of Pesticides in Soil-Water Systems: A Review of the Available Data Base.* Washington, DC: U.S. Environmental Protection Agency.

Overcash, Michael R. and James M. Davidson. (1980) *Environmental Impact of Nonpoint Source Pollution.* Ann Arbor, MI: Ann Arbor Science Press.

Pacenka, Steven and Tammo Steenhuis. (1984) *User's Guide for the MOUSE Computer Program.* Ithaca: Cornell University Dept. of Agr. Engr.

Page, R. Talbot (1973) *The Economics of Involuntary Transfers.* Berlin: Springer-Verlag.

―――. (1978) "A Generic View of Toxic Chemicals and Similar Risks." *Ecology Law Quarterly* 7: 207-244.

Papendick, R. I., L. F. Elliot and R. B. Dahlgren (1986) "Environmental Consequences of Modern Production Agriculture: How Can Alternative Agriculture Address These Concerns?" *American Journal of Alternative Agriculture* 1 (Winter): 3-10.

Papendick, R. I., L. F. Elliot and James F. Power. (1987) "Alternative Production Systems to Reduce Nitrates in Groundwater." *American Journal of Alternative Agriculture* 2 (Winter): 19-24.

Paris, Quirino and Gordon Rausser. (1973) "Sufficient Conditions for Aggregation of Linear Programming Models." *Am. J. Agr. Econ.* 55:659-666.

Perkins, John. () *Insects, Experts and the Insecticide Crisis.* New York: Plenum Press.

Peskin, Henry M. (1979) "Environmental Policy and the Distribution of Benefits and Costs." *Policy Study Annual Review.*

Petersen, C. and C. McCarthy. () "Farmland Preservation by Purchase of Development Rights: The Long Island Experiment." *DePaul L. Rev.* 26: 447.

Pfeiffer, George H. and Norman K. Whittlesey. (1978) "Controlling Nonpoint Externalities with Input Restrictions in an Irrigated River Basin." *Water Resources B.* 14(6) (December): 1387-1403.

Phelps, Joel B. and R. Brian How. (1981) Planning Data for Small Scale Commercial Vegetable and Strawberry Production in New York. A. E. Res. 81-20 Ithaca: Cornell University Dept. of Agricultural Economics.

Pimentel, David (1978) "Socioeconomic and Legal Aspects of Pest Control" in Edward H. Smith and David Pimentel (eds.) *Pest Control Strategies.* New York: Academic Press.

Pimentel, David and Christine Shoemaker. (1974) "An Economic and Land Use Model for Reducing Insecticide Use on Cotton and Corn." *Environmental Entomology.* 3(1): 10-20.

Porter, Keith S. (1978) Nitrogen on Long Island, Sources and Fate. Ithaca: Cornell University Cooperative Extension.

Porter, Keith, et. al. (1975) *Nitrogen and Phosphorus.* Ann Arbor: Ann Arbor Science Press.

Post, Henry. (1981) "Forget the Hamptons, Now It's Country Chic. *New York.* August 10: 24 ff.

Powledge, Fred. (1982) *Water.* New York: Farrar, Straus, Giroux.

Pye, Veronica I. Ruth Patrick and John Quarles. (1983) *Groundwater Contamination in the United States.* Philadelphia: University of Pennsylvania Press.

Rao, P. S. C. and J. M. Davidson. (1980) "Estimation of Pesticide Retention and Transformation Parameters Required in Nonpoint Source Pollution Models," in Overcash and Davidson,

Environmental Impact of Nonpoint Source Pollution. Ann Arbor, MI: Ann Arbor Science Press: 23-67.

————. (1982) *Transformation of Selected Pesticides and Phosphorus in Soil Water Systems: A Critical Review.* Springfield, VA: National Technical Information Service.

————. (1982) Retention and Transformation of Selected Pesticides and Phosphorus in Soil Water Systems: A Critical Review. Springfield, VA: National Technical Information Service.

Rao, P. S. C., A. G. Hornsby, and R. E. Jessup. (1985) "Indices for Ranking the Potential for Pesticide Contamination of Groundwater." *Soil Crop Sci. Soc. Florida Proc.* 44: (forthcoming).

Rao, P. S. C., R. S. Mansell, L. B. Baldwin and M. F. Laurent. (1983) "Pesticides and Their Behavior in Soil and Water." Gainesville, FL: U. Florida Dept. of Soil Science.

Rausser, Gordon C., Richard E. Just and David Zilberman. (1980) "Prospects and Limitations of Operations Research Applications in Agriculture and Agricultural Policy" in D. Yaron and C. S. Tapiero (eds.) *Operations Research in Agriculture and Water Resources*: 17-40. Amsterdam: North-Holland Publishing Company.

Raymond, Lyle S. (ed.) (1981) *Groundwater Management in the Northeastern Sates: Legal and Institutional Issues.* Ithaca: Center for Environmental Research.

Recker, Paula M. (1976) "Animal Feeding Factories and the Environment: A Summary of Feedlot Pollution, Federal Controls and Oklahoma Law." *Sw. Law J.* 30: 556ff.

Roberts, E. F. (1982) *The Law and the Preservation of Agricultural Land.* Ithaca: Center for Northeast Development.

Rossi, Daniel, Pritam S. Dillon and Laura Hoffman. (1983) "Apple Pest Management: A Cost Analysis of Alternative Practices." *Northeast J. Agr. Econ.* 12(2): 77-81.

Rovinsky, Robert. (1977) *Models to Evaluate the Economic and Environmental Effects of Restricting Pesticide Use on Cotton and Corn.* Ithaca: Cornell Univ. unpublished Ph. D. dissertation.

Russell, Clifford S. (ed.) (1978) *Safe Drinking Water: Current and Future Problems.* Washington, DC: Resources for the Future.

Rykbost, K. A., P. A. Schippers, R. S. Kossack, G. W. Selleck, C. C. Chu and D. Bouldin. (1978) *Studies in Fertility and Nitrate Pollution in Potatoes on Long Island.* Riverhead, NY: Long Island Horticultural Research Laboratory.

Sampson, R. Neil. (1981) *Farmland or Wasteland.* Emmaus, PA: Rodale Press.

Sarhan, Mohamed E., Richard E. Hewitt and Charles V. Moore. (1979) 1979 "Pesticide Resistance Externalities and Optimal Mosquito Management." *J. Environ. Econ. and Mgmt.*

Schaffer, William Harry. (1974) *An Economic Analysis of Nitrogen, Phosphorus and Soil Loss from Agricultural Production Affecting Water Quality in a Small New York Watershed.* Ithaca: Cornell Univ. unpublished Ph. D. dissertation.

Schaller, Neill W. and Gerald W. Dean. (1965) *Predicting Regional Crop Production: An Application of Recursive Programming.* U.S.D.A. Technical Bulletin no. 1329. Washington, DC: US Department of Agriculture.

Schelling, T. C. (1983) *Incentives for Environmental Protection* Cambridge, MA: MIT Press.

Schloesser, Lynn C. (1978) "Agricultural Non-Point Source Water Pollution Control under Sections 208 and 303 of the Clean Water

Act: Has Forty Years of Experience Taught Us Anything?" *N. Dak. Law R.* 54: 589-618.

Schmid, A. Allan. (1968) Converting Land from Rural to Urban Uses. Baltimore: Resources for the Future, Johns Hopkins Press.

Schmidt, Kenneth D. (1977) "Protection of Groundwater from Nonpoint Sources of Pollution." in Robert S. Pojasek (ed.) *Drinking Water Quality Enhancement Through Source Protection*: 257-273. Ann Arbor, MI: Ann Arbor Science Press.

Schnaiberg, A., N. Watts, and K. Zimmerman (1986) *Distributional Conflicts in Environmental Resource Policy*. New York: St. Martins Press.

Schneider, Keith. (1983) "The Case Against Industrial Bio-Test." *Amicus Journal* 4(4): 14-26

Scholvinck, Johan B. W. (1974) *Preserving Agriculture on Long Island*. Ithaca: Cornell University Unpublished M.S. Thesis.

Sharefkin, Mark F., Mordechai Schechter and Allen V. Kneese. (1983) "Impacts, Costs and Techniques for Mitigation of Contaminated Groundwater." *Groundwater Resources and Contamination in the United States*: 93-126. Washington, DC: National Science Foundation.

Sharples, Jerry A. (1969) "Predicting Short-Run Aggregate Adjustment to Policy Alternatives." *Am. J. Agr. Econ.* 51(2): 344-358.

Shih, C. S. and J. W. Ingam. (1981) "Multiattribute Decision Analysis for the Protection of Groundwater Resources." *J. Hydrology.* 51: 265-75

Shoemaker, Christine A. (1973) "Optimization of Agricultural Pest Management (Part I)." *Mathematical Biosciences.* 16: 143-175.

————. (1977) "Pest Management Models of Crop Ecosystems" in Charles A. S. Hall and John W. Day (eds.) *Ecosystem Modeling in Theory and Practice: An Introduction With Case Histories.* New York: John Wiley and Sons.

————. (1981) "Applications of Dynamic Programming and Other Optimization Methods in Pest Management." *IEEE Transactions on Automatic Control.* AC-26(5) (October): 1125-1132.

Shroder, R. F. W. and M. M. Athanas (1985) "Review of Research on *Edovum putlerri* Grissell, Egg Parasite of the Colorado Potato Beetle" In *Proceedings of the Symposium on the Colorado Potato Beetle, XVIIth International Congress of Entomology,* D. N. Ferro and R. H. Voss (eds.) Massachussetts Experiment Station, Amherst.

Simon, Herbert A. (1955) "A Behavioral Model of Rational Choice." *Q. J. Econ.* 69(1) (February): 99-118.

————. (1957) *Administrative Behavior.* New York: Macmillan Press.

Slovic, Paul, Baruch Fischhoff and Sarah Lichtenstein. (1982) "Facts versus Fears: Understanding Perceived Risk." Kahneman, Slovic, Tversky (eds.) *Judgement Under Uncertainty:Heuristics and Biases.* Cambridge: Cambridge University Press.

Smith, V. Kerry. (1987) "Uncertainty, Benefit Cost Analysis and the Treatment of an Option Value." *Journal of Environmental Economics and Management* 14: 283-292.

Snyder, D. P. (1977) "Cost of Production Update for 1976--Long Island Potatoes." Ithaca: Cornell University Dept. of Agr. Econ. A. E. Ext. 77-11.

————. (1982) "Cost of Production Update for 1981--Long Island Potatoes." Ithaca: Cornell University Dept. of Agr. Econ. A. E. Ext. 82-20.

————. (1983) "Cost of Production Update for 1982--Long Island Cauliflower." Ithaca: Cornell University Dept. of Agr. Econ. A. E. Ext. 83-30.

————. (1983) *Farm Cost Accounts*. Ithaca: Cornell University Dept. of Agr. Econ. A. E. Ext. 83-41.

Spengler, Joseph J. (1976) "Limits to Growth: Biospheric or Institutional?" in William Breit and William Patton Culbertson, Jr., (eds.) *Science and Ceremony*: 115-133. Austin: Univ. of Texas Press.

Stanton, B. F. and W. A. Knoblauch. (1984) "New York Agriculture Census Data, 1982." Ithaca: Cornell Univ. Dept. of Agricultural Economics, A. E. Ext. 84-21

Starr, Chauncey and Chris Whipple. (1980) "Risks of Risk Decisions." *Science*. 208: 1114-1119.

Stevie, Richard G., Robert M. Clark, Jeffrey Q. Adams and James I. Gillean. (1979) *Managing Small Water Systems*. Cincinnati: U.S. Environmental Protection Agency.

Stokey, Edith and Richard Zeckhauser (1978) *A Primer for Policy Analysis*. New York: W. W. Norton

Strobehn, Roger W., Beatrice H. Holmes, William D. Anderson and Charles D. Reese. (1973) *Laws and Institutional Mechanisms Controlling the Release of Pesticides into the Environment*. Washington, DC: US Environmental Protection Agency and US Department of Agriculture.

Suffolk County Cooperative Extension. (1983) "Facts About Suffolk's Agricultural Industry." Riverhead, NY: Cooperative Extension.

Swartzman, Daniel, Richard A. Liroff and Kevin G. Croke. (1982) *Cost Benefit Analysis and Environmental Regulations: Politics, Ethics and Methods*. Washington, DC: Conservation Foundation.

Thomson, W. T. (1990) *Agricultural Chemicals*. Fresno, CA: Thomson Publications.

Thurow, Lester. (1980) *The Zero-Sum Society*. New York: Basic Books.

Tinsley, Ian J. (1979) *Chemical Concepts in Pollution Behavior*. New York: John-Wiley Interscience.

Trautman, Nancy M., Keith S. Porter and Henry B. F. Hughes. (1983) "Protection of Restoration of Groundwater in Southold, N.Y." Ithaca: Cornell University Center for Environmental Research.

————. (1983) "Southold Demonstration Site." Ithaca: Cornell University Center for Environmental Research.

Trelease, Frank J. (1979) *Water Law: Cases and Materials*. St. Paul: West Publishing Co.

Tripp, J. T. B. and A. Jaffe. (1979) "Preventing Groundwater Pollution: Towards a Coordinated Strategy to Protect Critical Recharge Zones." *Harvard Environ. Law Rev.* 3: 364-410.

Tversky, Amos and Daniel Kahneman. (1974) "Judgement Under Uncertainty." *Science*. 185: 1124-1131.

U.S. Congress. House. Committee on Agriculture. Subcommittee on Department Operations, Research and Foreign Agriculture. (1983) *Regulation of Pesticides* v. IV Washington, DC: U.S. Government Printing Office.

U.S. Congress, Office of Technology Assessment. (1984) *Protecting the Nation's Groundwater From Contamination*. Washington, DC: OTA Reports.

U.S. Dept. of Agriculture. (1975) *Soil Survey of Suffolk County, NY*. Washington, DC: U.S. Government Printing Office.

U. S. Environmental Protection Agency. (Various Years) *Pesticide Abstracts*. Washington, DC: U.S. Government Printing Office.

————. (1986) *National Survey of Pesticides in Drinking Water Wells*. Washington: US Environmental Protection Agency

U.S. General Accounting Office (1979) *Federal Pesticide Registration Program: Is it Protecting the Public and the Environment Adequately from Pesticide Hazards?* Washington, DC: US Government Printing Office.

————. (1979) *Improving the Scientific and Technical Information Available to the Environmental Protection Agency in its Decisionmaking Process*. Washington: U.S. Government Printing Office.

————. (1982) *States Compliance Lacking in Meeting Safe Drinking Water Regulations*. Washington, DC: U.S. Government Printing Office.

————. (1984) *Cost-Benefit Analysis Can Be Useful in Assessing Environmental Regulations, Despite Limitations*. Washington, DC: U.S. Government Printing Office.

U. S. Library of Congress. Congressional Research Service. Environmental Policy Division. (1975) *Potential Effects of Application of Air and Water Quality Standards on Agriculture and Rural Development*. Washington, DC: U. S. Government Printing Office.

————. (1979) *Agricultural and Environmental Relationships: Issues and Priorities*. Washington, DC: US Government Printing Office.

————. (1982) *A Legislative History of the Clean Water Act*. Washington, DC: U.S. Congressional Research Service.

Underwood, Floyd W. (1932) *Factors Affecting Costs and REturns in Producing Potatoes in New York in 1929*. Ithaca: Cornell University Unpublished Ph.D. Dissertation.

van den Bosch, Robert. (1978) *The Pesticide Conspiracy*. Garden City, NY: Doubleday Co.

Vaupel, James W. (1982) "Truth and Consequences: Some Roles for Scientists and Analysts in Environmental Decisionmaking." in Wesley A. Magat (ed.) *Reform of Environmental Regulation*: 71-92. Cambridge, MA: Ballinger Publishing.

Varian, Hal R. (1983) *Microeconomic Analysis*. New York: W.W. Norton, Inc.

Wagenet, L. P., A. T. Lemley and R. J. Wagenet. (1985) "A Review of Physical Chemical Parameters Related to the Soil and Groundwater Fate of Selected Pesticides Used in New York State." Ithaca: Cornell University.

Warner, Mildred E. (1985) *Alternatives for Long Island Agriculture: The Economic Potential of Peaches and Table Grapes*. Ithaca: Cornell University unpublished M.S. Thesis.

Watson, David L. and A. W. A. Brown. (1977) *Pesticide Management and Insecticide Resistance*. New York: Academic Press.

Watson, William D. and Ronald G. Ridker. (1981) "Revising Water Pollution Standards in an Uncertain World." *Land Econ.* 57(4): 485-506.

Weber, J. B. (1975) "Agricultural Chemicals and Their Importance as a Non-Point Source of Water Pollution," in Ashton and Underwood, *Non-Point Sources of Water Pollution*. Blacksburg: Virginia Polytechnic Institute and State University, Virginia Water Resources Center.

Weidenbaum, Murray. (1978) *The Cost of Federal Regulation of Economic Activities*. Washington, DC: American Enterprise Institute.

Weigold, Marilyn. (1974) *American Mediterranean.* Port Washington, NY: Kennikat Press.

White, Gerald, David Lee and Ken Gardner. (1983) "Survey of Farm Businesses in Suffolk County." *Suffolk County Agricultural News.* (September): 3-4.

White, Gerald B. and Jeannette L. Smith. (1986) Cost of Production for *Vinifera* Grapes on Long Island, 1985. Ithaca, NY: Cornell University Dept. of Agi. Econ., A.E. Res. 86-11.

Whitten, Jamie L. (1966) *That We May Live.* Princeton, NJ: D. van Nostrand Co.

Wildavsky, Aaron. (1980) "Richer is Safer." *Public Interest.* 60: 23-39.

Windholz, Martha (ed.). (1983) *The Merck Index.* Rahway, NJ: Merck & Co., Inc.

Worthing, Charles R. (ed.). (1983) *The Pesticide Manual.* Lavenham, England: British Crop Protection Council, Lavenham Press, Ltd.

Wright, R. J., R. Loria, J. B. Sieczka and D. D. Moyer. (1984) Final report of the 1983 Long Island integrated pest management pilot program. Cornell University Dept. of Veg. Crops, V. C. Mimeo, Ithaca, NY.

Wynne, Brian. (1982) *Rationality and Ritual.* Chalfant St. Giles, UK: British Society for the History of Science.

Yaron, D. and C. S. Tapiero (eds.) (1980) *Operations Research in Agriculture and Water Resources.* Amsterdam: North-Holland Publishing Company.

Zimmerman, Rae. (1979) "Institutional Constraints on Land Management for Water Resource Protection in Urban and Suburban Watersheds." New York: New York University Public Policy Research Institute.

THE ENVIRONMENT

A. STANLEY MEIBURG. *Project and Enhance: "Juridical Democracy" and the Prevention of Significant Deterioration of Air Quality.*

A. RICHARD MORDI. *Attitudes Toward Wildlife in Botswana.*

ROBERT W. MORESCHI. *Tort Liability Standards and the Firm's Response to Regulation.*

USHA MUNSHI. *An Integrated Approach to Pollution Control.*

RICHARD B. OLOWOMEYE. *The Management of Solid Waste in Nigerian Cities.*

COLLEEN K. O'TOOLE. *The Search for Purity: A Retrospective Policy Analysis of the Decision to Chlorinate Cincinnati's Public Water Supply, 1890–1920.*

ROBERT B. OLSHANSKY. *Landslide Hazard in the United States: Case Studies in Planning and Policy Development.*

KENNETH B. SEWALL. *The Tradeoff Between Cost and Risk in Hazardous Waste Management.*

SMITA K. SIDDHANTI. *Multiple Perspectives on Risk and Regulation: The Case of Deliberate Release of Genetically Engineered Organisms into the Environment.*

ROBERT N. STAVINS. *The Welfare Economics of Alternative Renewable Resource Strategies: Forested Wetlands and Agricultural Production.*

ELAINE VAUGHAN. *Some Factors Influencing the Nonexperts Perception and Evaluation of Environmental Risks.*

LEAH J. WILDS. *Understanding Who Wins: Organizational Behavior and Environmental Policies.*

DAVID K. WOODYARD AND JONATHAN HAUFLER. *Risk Evaluation for Sludge-Born Elements to Wildlife Food Chains.*